Experiments and Observations on Different Kinds of Air

VOLUME 1

JOSEPH PRIESTLEY

CAMBRIDGE
UNIVERSITY PRESS

CAMBRIDGE
UNIVERSITY PRESS

University Printing House, Cambridge, CB2 8BS, United Kingdom

Published in the United States of America by Cambridge University Press, New York

Cambridge University Press is part of the University of Cambridge.
It furthers the University's mission by disseminating knowledge in the pursuit of
education, learning and research at the highest international levels of excellence.

www.cambridge.org
Information on this title: www.cambridge.org/9781108063951

This edition first published 1775
This digitally printed version 2013

ISBN 978-1-108-06395-1 Paperback

CAMBRIDGE LIBRARY COLLECTION

Books of enduring scholarly value

Physical Sciences

From ancient times, humans have tried to understand the workings of the world around them. The roots of modern physical science go back to the very earliest mechanical devices such as levers and rollers, the mixing of paints and dyes, and the importance of the heavenly bodies in early religious observance and navigation. The physical sciences as we know them today began to emerge as independent academic subjects during the early modern period, in the work of Newton and other 'natural philosophers', and numerous sub-disciplines developed during the centuries that followed. This part of the Cambridge Library Collection is devoted to landmark publications in this area which will be of interest to historians of science concerned with individual scientists, particular discoveries, and advances in scientific method, or with the establishment and development of scientific institutions around the world.

Experiments and Observations on Different Kinds of Air

By the late eighteenth century, scientists had discovered certain types of gas, such as 'fixed air' (carbon dioxide), but their composition was little understood. Relatively few investigations into gases had taken place, and so the polymath Joseph Priestley (1733–1804) was able to make major breakthroughs in the field using a range of experimental techniques. While living near a brewery, he found that it was possible to outline the shape of the gas above fermenting beer with smoke, and that fire would burn with varying strength depending on the composition of the air. This three-volume collection first appeared between 1774 and 1777. Primarily an account of Priestley's early experiments, with details of apparatus including candles and live mice, Volume 1 is reissued here in its corrected 1775 second edition and also incorporates a brief history of the field of inquiry.

Cambridge University Press has long been a pioneer in the reissuing of out-of-print titles from its own backlist, producing digital reprints of books that are still sought after by scholars and students but could not be reprinted economically using traditional technology. The Cambridge Library Collection extends this activity to a wider range of books which are still of importance to researchers and professionals, either for the source material they contain, or as landmarks in the history of their academic discipline.

Drawing from the world-renowned collections in the Cambridge University Library and other partner libraries, and guided by the advice of experts in each subject area, Cambridge University Press is using state-of-the-art scanning machines in its own Printing House to capture the content of each book selected for inclusion. The files are processed to give a consistently clear, crisp image, and the books finished to the high quality standard for which the Press is recognised around the world. The latest print-on-demand technology ensures that the books will remain available indefinitely, and that orders for single or multiple copies can quickly be supplied.

The Cambridge Library Collection brings back to life books of enduring scholarly value (including out-of-copyright works originally issued by other publishers) across a wide range of disciplines in the humanities and social sciences and in science and technology.

EXPERIMENTS

AND

OBSERVATIONS

ON DIFFERENT KINDS OF

A I R.

By JOSEPH PRIESTLEY, L L. D. F. R. S.

The SECOND EDITION Corrected.

Fert animus Caufas tantarum expromere rerum ;
Immenfumque aperitur opus.

LUCAN

L O N D O N:

Printed for J. JOHNSON, No. 72, in St. Paul's
Church-Yard.

MDCCLXXV.

TO THE RIGHT HONOURABLE

THE EARL OF SHELBURNE,

THIS TREATISE IS

WITH THE GREATEST GRATITUDE

AND RESPECT,

INSCRIBED,

BY HIS LORDSHIP's

MOST OBLIGED,

AND OBEDIENT

HUMBLE SERVANT,

J. PRIESTLEY.

THE

PREFACE.

ONE reason for the present publication has been the favourable reception of those of my *Observations on different kinds of air*, which were published in the Philosophical Transactions for the year 1772, and the demand for them by persons who did not chuse, for the sake of those papers only, to purchase the whole volume in which they were contained. Another motive was the *additions* to my observations on this subject, in consequence of which my papers grew too large for such a publication as the *Philosophical Transactions*.

Contrary, therefore, to my intention, expressed Philosophical Transactions, vol. 64. p. 90, but with the approbation of the President, and of my friends in the so-

ciety

ciety, I have determined to fend them no
more papers for the prefent on this fubject,
but to make a feparate and immediate
publication of all that I have done with
refpect to it.

Befides, confidering the attention which,
I am informed, is now given to this fubject by philofophers in all parts of Eu-
rope, and the rapid progrefs that has already been made, and may be expected to
be made in this branch of knowledge, all
unneceffary delays in the publication of
experiments relating to it are peculiarly
unjuftifiable.

When, for the fake of a little more reputation, men can keep brooding over a
new fact, in the difcovery of which they
might, poffibly, have very little real merit,
till they think they can aftonifh the world
with a fyftem as complete as it is new,
and give mankind a prodigious idea of
their judgment and penetration; they are
juftly punifhed for their ingratitude to the
fountain of all knowledge, and for their
want of a genuine love of fcience and of
man-

mankind, in finding their boafted difcoveries anticipated, and the field of honeft
fame pre-occupied, by men, who, from a
natural ardour of mind, engage in philofophical purfuits, and with an ingenuous
fimplicity immediately communicate to
others whatever occurs to them in their
inquiries.

As to myfelf, I find it abfolutely impoffible to produce a work on this fubject
that fhall be any thing like *complete.* My
firft publication I acknowledged to be very
imperfect, and the prefent, I am as ready
to acknowledge, is ftill more fo. But, pa-
radoxical as it may feem, this will ever be
the cafe in the progrefs of natural fcience,
fo long as the works of God are, like himfelf, infinite and inexhauftible. In com-
pleting one difcovery we never fail to get
an imperfect knowledge of others, of
which we could have no idea before; fo
that we cannot folve one doubt without
creating feveral new ones.

Travelling on this ground refembles
Pope's defcription of travelling among

the Alps, with this difference, that here there is not only a *succeſſion*, but an *increaſe* of new objects and new difficulties.

> So pleas'd at firſt the tow'ring Alps we try,
> Mount o'er the vales, and ſeem to tread the ſky.
> Th' eternal ſnows appear already paſt,
> And the firſt clouds and mountains ſeem the laſt,
> But thoſe attain'd, we tremble to ſurvey
> The growing labours of the lengthen'd way.
> Th' increaſing proſpect tires our wand'ring eyes,
> Hills peep o'er hills, and Alps on Alps ariſe.
> Essay on Criticism.

Newton, as he had very little knowledge of *air*, ſo he had few doubts concerning it.   Had Dr. Hales, after his various and valuable inveſtigations, given a liſt of all his *deſiderata*, I am confident that he would not have thought of one in ten that had occurred to me at the time of my laſt publication ; and my doubts, queries, and hints for new experiments are very conſiderably increaſed, after a ſeries of inveſtigations, which have thrown great light upon many things of which I was not able to give any explanation before.

I would

I would obferve farther, that a perfon who means to ferve the caufe of fcience effectually, muft hazard his own reputation fo far as to rifk even *miftakes* in things of lefs moment. Among a multiplicity of new objects, and new relations, fome will neceffarily pafs without fufficient attention; but if a man be not miftaken in the principal objects of his purfuits, he has no occafion to diftrefs himfelf about leffer things.

In the progrefs of his inquiries he will generally be able to rectify his own miftakes ; or if little and envious fouls fhould take a malignant pleafure in detecting them for him, and endeavouring to expofe him, he is not worthy of the name of a philofopher, if he has not ftrength of mind fufficient to enable him not to be difturbed at it. He who does not foolifhly affect to be above the failings of humanity, will not be mortified when it is proved that he is but a man.

In this work, as well as in all my other philofophical writings, I have made it a

rule

rule not to conceal the *real views* with
which I have made experiments ; becaufe
though, by following a contrary maxim,
I might have acquired a character of greater fagacity, I think that two very good
ends are anfwered by the method that I
have adopted. For it both tends to make a
narrative of a courfe of experiments more
interefting, and likewife encourages other
adventurers in experimental philofophy ;
fhewing them that, by purfuing even falfe
lights, real and important truths may be
difcovered, and that in feeking one thing
we often find another.

In fome refpects, indeed, this method
makes the narrative *longer*, but it is by
making it lefs tedious; and in other refpects I have written much more concifely
than is ufual with thofe who publifh accounts of their experiments. In this treatife the reader will often find the refult of
long proceffes expreffed in a few lines, and
of many fuch in a fingle paragraph ;
each of which, if I had, with the ufual
parade, defcribed it at large (explaining
firft the *preparation*, then reciting the *ex-*
periment

periment itſelf, with the *reſult* of it, and laſtly making ſuitable *reflections*) would have made as many ſections or chapters, and have ſwelled my book to a pompous and reſpectable ſize. But I have the plea-ſure to think that thoſe philoſophers who have but little time to ſpare for *reading*, which is always the caſe with thoſe who *do* much themſelves, will thank me for not keeping them too long from their own purſuits ; and that they will find ra-ther more in the volume, than the appear-ance of it promiſes.

I do not think it at all degrading to the buſineſs of experimental philoſophy, to compare it, as I often do, to the diver-ſion of *hunting*, where it ſometimes hap-pens that thoſe who have beat the ground the moſt, and are conſequently the beſt acquainted with it, weary themſelves without ſtarting any game ; when it may fall in the way of a mere paſſenger ; ſo that there is but little room for boaſting in the moſt ſucceſsful termination of the chace.

The

The beſt founded praiſe is that which is due to the man, who, from a ſupreme veneration for the God of nature, takes pleaſure in contemplating his *works*, and from a love of his fellow-creatures, as the offspring of the ſame all-wiſe and benevolent parent, with a grateful ſenſe and perfect enjoyment of the means of happineſs of which he is already poſſeſſed, ſeeks, with earneſtneſs, but without murmuring or impatience, that greater *command of the powers of nature*, which can only be obtained by a more extenſive and more accurate *knowledge* of them; and which alone can enable us to avail ourſelves of the numerous advantages with which we are ſurrounded, and contribute to make our common ſituation more ſecure and happy.

Beſides, the man who believes that there is a *governor* as well as a *maker* of the world (and there is certainly equal reaſon to believe both) will acknowledge his providence and favour at leaſt as much

in

in a fuccefsful purfuit of *knowledge*, as of *wealth*; which is a fentiment that entirely cuts off all boafting with refpect to ourfelves, and all envy and jealoufy with refpect to others; and difpofes us mutually to rejoice in every new light that we receive, through whofe hands foever it be conveyed to us.

I fhall pafs for an enthufiaft with fome, but I am perfectly eafy under the imputation, becaufe I am happy in thofe views which fubject me to it; but confidering the amazing improvements in natural knowledge which have been made within the laft century, and the many ages, abounding with men who had no other object but ftudy, in which, however, nothing of this kind was done, there appears to me to be a very particular providence in the concurrence of thofe circumftances which have produced fo great a change; and I cannot help flattering myfelf that this will be inftrumental in bringing about other changes in the ftate

of

of the world, of much more confequence to the improvement and happinefs of it.

This rapid progrefs of knowledge, which, like the progrefs of a *wave* of the fea, of *found*, or of *light* from the fun, extends itfelf not this way or that way only, but *in all directions*, will, I doubt not, be the means, under God, of extirpating *all* error and prejudice, and of putting an end to all undue and ufurped authority in the bufinefs of *religion*, as well as of *fcience*; and all the efforts of the interefted friends of corrupt eftablifhments of all kinds will be ineffectual for their fupport in this enlightened age: though, by retarding their downfal, they may make the final ruin of them more complete and glorious. It was ill policy in Leo the Xth to patronize polite literature. He was cherifhing an enemy in difguife. And the Englifh hierarchy (if there be any thing unfound in its conftitution) has equal reafon to tremble even at an air-pump, or an electrical machine.

There

There certainly never was any period
in which *natural knowledge* made fuch
a progrefs as it has done of late years,
and efpecially in this country; and they
who affect to fpeak with fupercilious con-
tempt of the publications of the prefent age in general, or of the Royal Society in particular, are only thofe who
are themfelves engaged in the moft trifling of all literary purfuits, who are
unacquainted with all real fcience, and
are ignorant of the progrefs and prefent ftate of it.*

It is true that the rich and the great
in this country give lefs attention to
thefe fubjects than, I believe, they were
ever known to do, fince the time of
Lord Bacon, and much lefs than men
of rank and fortune in other countries
give to them. But with us this lofs is

* See Sir John Pringle's *Difcourfe on the different kinds
of air*, p. 29, which, if it became me to do it, I
would recommend to the reader, as containing a juft
and elegant account of the feveral difcoveries that
have been fucceffively made, relating to the fubject of
this treatife.

I   made

made up by men of leifure, fpirit, and
ingenuity, in the middle ranks of life,
which is a circumftance that promifes
better for the continuance of this progrefs in ufeful knowledge than any no-
ble or royal patronage. With us, politics
chiefly engage the attention of thofe who
ftand foremoft in the community, which,
indeed, arifes from the *freedom* and peculiar *excellence* of our conftitution, without which even the fpirit of men
of letters in general, and of philofophers in particular, who never directly
interfere in matters of government, would
languifh.

It is rather to be regretted, however,
that, in fuch a number of nobility and
gentry, fo very few fhould have any tafte
for fcientifical purfuits, becaufe, for many
valuable purpofes of fcience, *wealth* gives
a decifive advantage. If extenfive and
lafting *fame* be at all an object, literary,
and efpecially fcientifical purfuits, are
preferable to political ones in a variety
of refpects. The former are as much
more favourable for the difplay of the

2 human

human faculties than the latter, as the *fyftem of nature* is fuperior to any *political fyftem* upon earth.

If extenfive *ufefulnefs* be the object, fcience has the fame advantage over politics. The greateft fuccefs in the latter feldom extends farther than one particular country, and one particular age; whereas a fuccefsful purfuit of fcience makes a man the benefactor of all mankind, and of every age. How trifling is the fame of any ftatefman that this country has ever produced to that of Lord Bacon, of Newton, or of Boyle; and how much greater are our obligations to fuch men as thefe, than to any other in the whole *Biographia Britannica*; and every country, in which fcience has flourifhed, can furnifh inftances for fimilar obfervations.

Here my reader will thank me, and the writer will, I hope, forgive me, if I quote a paffage from the poftfcript of a letter which I happen to have juft received from that excellent, and in my

b   opi-

xviii PREFACE.

opinion, not too enthufiaftical philofopher, father Beccaria of Turin.

*Mi fpiace che il mondo politico ch' è pur
tanto paffeggero, rubbi il grande Franklin
al mondo della natura, che non fa ne cam-
biare, ne mancare.* In Englifh. " I am
" forry that the *political world*, which is
" fo very tranfitory, fhould take the
" great Franklin from the *world of na-*
" *ture*, which can never change, or fail."

I own it is with peculiar pleafure that
I quote this paffage, refpecting this truly
great man, at a time when fome of the
infatuated politicians of this country are
vainly thinking to build their wretched
and deftructive projects, on the ruins of
his eftablifhed reputation; a reputation
as extenfive as the fpread of fcience itfelf, and of which it is faying very little indeed, to pronounce that it will laft
and flourifh when the names of all his
enemies fhall be forgotten.

I think

I think it proper, upon this occasion, to inform my friends, and the public, that I have, for the present, suspended my design of writing *the history and pre-sent state of all the branches of experi-mental philosophy.* This has arisen not from any dislike of the undertaking, but, in truth, because I see no prospect of be-ing reasonably indemnified for so much labour and expence, notwithstanding the specimens I have already given of that work (in the *history of electricity*, and of the *discoveries relating to vision, light, and colours*) have met with a much more favourable reception from the best judges both at home and abroad, than I expected. Immortality, if I should have any view to it, is not the proper price of such works as these.

I propose, however, having given so much attention to the subject of *air*, to write, at my leisure, the history and

present

prefent ftate of difcoveries relating to it;
in which cafe I fhall, as a part of it, re-
print this work, with fuch improvements
as fhall have occurred to me at that time;
and I give this notice of it, that no perfon who intends to purchafe it may have
reafon (being thus apprifed of my intention) to complain of buying the fame
thing twice. If any perfon chufe it, he
may fave his five or fix fhillings for the
prefent, and wait five or fix years longer
(if I fhould live fo long) for the opportunity of buying the fame thing, probably much enlarged, and at the fame
time a complete account of all that has
been done by others relating to this fubject.

Though for the plain, and I hope fatisfactory reafon above mentioned, I fhall
probably write no other *hiftories* of this
kind, I fhall, as opportunity ferves, endeavour to provide *materials* for fuch hiftories, by continuing my experiments,
keeping my eyes open to fuch new appearances as may prefent themfelves, inveftigating them as far as I fhall be able, and
never

never failing to communicate to the public, by fome channel or other, the refult
of my obfervations.

In the publication of this work I have
thought that it would be agreeable to my
readers to preferve, in fome meafure, the
order of hiftory, and therefore I have not
thrown together all that I have obferved
with refpect to each kind of air, but have
divided the work into *two parts* ; the former containing what was publifhed before,
in the Philofophical Tranfactions, with
fuch obfervations and corrections as fubfequent experience has fuggefted to me;
and I have referved for the latter part of
the work an account of the experiments
which I have made fince that publication,
and after a pretty long interruption in
my philofophical purfuits, in the courfe of
the laft fummer. Befides I am fenfible
that in the latter part of this work a different arrangement of the fubjects will be
more convenient, for their mutual il-
luftration.

Some

Some perfons object to the term *air*, as applied to *acid, alkaline,* and even *nitrous air* ; but it is certainly very convenient to have a common term by which to denote things which have fo many common properties, and thofe fo very ftriking ; all of them agreeing with the air in which we breathe, and with *fixed air*, in *elafticity,* and *tranfparency*, and in being alike affected by heat or cold ; fo that to the eye they appear to have no difference at all. With much more reafon, as it appears to me, might a perfon object to the common term *metal*, as applied to things fo very different from one another as gold, quickfilver, and lead.

Befides, *acid* and *alkaline* air do not differ from *common air* (in any refpect that can countenance an objection to their having a common appellation) except in fuch properties as are common to it with *fixed air*, though in a different degree ; viz. that of being imbibed by water. But, indeed, all kinds of air, common air itfelf not excepted, are capable of being imbibed by water in fome degree.

Some

Some may think the terms acid and alkaline *vapour* more proper than acid and
alkaline *air*. But the term *vapour* having
always been applied to elaftic matters capable of being condenfed in the temperature of the atmofphere, efpecially the vapour of water, it feems harfh to apply it
to any elaftic fubftance, which at the fame
time that it is as tranfparent as the air
we breathe, is no more affeted by cold
than it is.

As my former papers were immediately
tranflated into feveral foreign languages,
I may prefume that this treatife, having a
better title to it, will be tranflated alfo ;
and, upon this prefumption, I cannot
help expreffing a wifh, that it may be done
by perfons who have a competent knowledge of the *fubjet*, as well as of the *Englifh language*. The miftakes made by fome
foreigners, have induced me to give this
caution.

London, Feb.
1774.

A D V E R-

ADVERTISEMENT.

THE *weights* mentioned in the courfe of this treatife are *Troy*, and what is called *an ounce meafure of air*, is the fpace occupied by an ounce weight of water, which is equal to 480 grains, and is, therefore, almoft two *cubic inches* of water; for one cubic inch weighs 254 grains.

THE

INTRODUCTION.

SECTION I.

A general view of PRECEDING DISCOVERIES *relating to air.*

FOR the better underftanding of the experiments and obfervations on different kinds of air contained in this treatife, it will be ufeful to thofe who are not acquainted with the hiftory of this branch of natural philofophy, to be informed of thofe facts which had been difcovered by others, before I turned my thoughts to the fubject ; which fuggefted, and by the help of which I was enabled to purfue, my inquiries. Let it be obferved, however, that I do not profefs to recite in this place *all* that had been difcovered concerning air, but only thofe difcoveries the knowledge of which is neceffary, in order to underftand what I have done myfelf ; fo that any perfon who is only acquainted with the general principles of natural philofophy, may be able to

B   read

read this treatife, and, with proper attention,
to underftand every part of it.

That the air which conftitutes the atmofphere in which we live has *weight*, and that
it is *elaſtic*, or confiſts of a compreffible and
dilatable fluid, were fome of the earlieſt difcoveries that were made after the dawning of
philofophy in this weſtern part of the world.

That elaſtic fluids, differing effentially
from the air of the atmofphere, but agreeing
with it in the properties of weight, elaſticity,
and tranfparency, might be generated from
folid fubſtances, was difcovered by Mr. Boyle,
though two remarkable kinds of factitious air,
at leaſt the effects of them, had been known
long before to all miners. One of thefe is
heavier than common air. It lies at the bottom
of pits, extinguiſhes candles, and kills animals
that breathe it, on which account it had obtained the name of the *choke damp*. The other
is lighter than common air, taking its place
near the roofs of fubterraneous places, and becaufe it is liable to take fire, and explode, like
gunpowder, it had been called the *fire damp*.
The word *damp* fignifies *vapour* or *exhalation*
in the German and Saxon language.

Though

Though the former of thefe kinds of air had been known to be noxious, the latter I believe had not been difcovered to be fo, having always been found in its natural ftate, fo much diluted with common air, as to be breathed with fafety. Air of the former kind, befides having been difcovered in various caverns, particularly the *grotta del Cane* in Italy, had alfo been obferved on the furface of fermenting liquors, and had been called *gas* (which is the fame with *geift*, or *fpirit*) by Van Helmont, and other German chymifts; but afterwards it obtained the name of *fixed air*, efpecially after it had been difcovered by Dr. Black of Edinburgh to exift, in a fixed ftate, in alkaline falts, chalk, and other calcareous fubftances.

This excellent philofopher difcovered that it is the prefence of the fixed air in thefe fubftances that renders them *mild*, and that when they are deprived of it, by the force of fire, or any other procefs, they are in that ftate which had been called *cauftic*, from their corroding or burning animal and vegetable fubftances.

Fixed air had been difcovered by Dr. Macbride of Dublin, after an obfervation of Sir John Pringle's, which led to it, to be in a confiderable degree antifeptic; and fince it is

B 2   extracted

extracted in great plenty from fermenting vegetables, he had recommended the ufe of *wort*
(that is an infufion of malt in water) as what
would probably give relief in the fea fcurvy,
which is faid to be a putrid difeafe.

Dr Brownrigg had alfo difcovered that the
fame fpecies of air is contained in great quantities in the water of the Pyrmont fpring at Spa
in Germany, and in other mineral waters, which
have what is called an *acidulous* tafte, and that
their peculiar flavour, brifknefs, and medicinal
virtues, are derived from this ingredient.

Dr. Hales, without feeming to imagine that
there was any material difference between
thefe kinds of air and common air, obferved
that certain fubftances and operations *generate*
air, and others *abforb* it ; imagining that the diminution of air was fimply a taking away
from the common mafs, without any alteration
in the properties of what remained. His experiments, however, are fo numerous, and various, that they are juftly efteemed to be the
folid foundation of all our knowledge of this
fubject.

Mr. Cavendifh had exactly afcertained the
fpecific gravities of fixed and inflammable air,
fhewing the former of them to be $1\frac{1}{2}$ heavier
than

than common air, and the latter ten times lighter. He alfo fhewed that water would imbibe more than its own bulk of fixed air.

Laftly, Mr. Lane difcovered that water thus impregnated with fixed air will diffolve a confiderable quantity of iron, and thereby become a ftrong chalybeate.

Thefe, I would obferve, are by no means all the difcoveries concerning air that have been made by the gentlemen whofe names I have mentioned, and ftill lefs are they all that have been made by others; but they comprife all the previous knowledge of this fubject that is neceffary to the underftanding of this treatife; except a few particulars, which will be mentioned in the courfe of the work, and which it is, therefore, unneceffary to recite in this place.

SECT.

SECTION II.

An account of the APPARATUS *with which the following experiments were made.*

RAther than defcribe at large the manner in which every particular experiment that I fhall have occafion to recite was made, which would both be very tedious, and require an unneceffary multiplicity of drawings, I think it more advifeable to give, at one view, an account of all my apparatus and inftruments, or at leaft of every thing that can require a defcription, and of all the different operations and proceffes in which I employ them.

It will be feen that my apparatus for experiments on air is, in fact, nothing more than the apparatus of Dr. Hales, Dr. Brownrigg, and Mr. Cavendifh, diverfified, and made a little more fimple. Yet notwithftanding the fimplicity of this apparatus, and the eafe with which all the operations are conducted, I would not have any perfon, who is altogether without experience, to imagine that he fhall be able to felect any of the following experiments, and immediately perform it, without difficulty or

blunder-

blundering. It is known to all perfons who are converfant in experimental philofophy, that there are many little attentions and precautions neceffary to be obferved in the conducting of experiments, which cannot well be defcribed in words, but which it is needlefs to defcribe, fince practice will neceffarily fuggeft them; though, like all other arts in which the hands and fingers are made ufe of, it is only *much practice* that can enable a perfon to go through complex experiments, of this or any other kind, with eafe and readinefs.

For experiments in which air will bear to be confined by water, I firft ufed an oblong trough made of earthen ware, as *a* fig. 1. about eight inches deep, at one end of which I put thin flat ftones, *b. b.* about an inch, or half an inch, under the water, ufing more or fewer of them according to the quantity of water in the trough. But I have fince found it more convenient to ufe a larger wooden trough, of the fame general fhape, eleven inches deep, two feet long, and 1½ wide, with a fhelf about an inch lower than the top, inftead of the flat ftones abovementioned. This trough being larger than the former, I have no occafion to make provifion for the water being higher or lower, the bulk of a jar or two not making fo great a difference as it did before.

The

The feveral kinds of air I ufually keep in *cylindrical jars*, as *c, c*, fig. 1, about ten inches long, and 2 ¼ wide, being fuch as I have generally ufed for electrical batteries, but I have likewife veffels of very different forms and fizes, adapted to particular experiments.

When I want to remove veffels of air from the large trough, I place them in *pots* or *difhes*, of various fizes, to hold more or lefs water, according to the time that I have occafion to keep the air, as fig. 2. Thefe I plunge in water, and flide the jars into them ; after which they may be taken out together, and be fet wherever it fhall be moft convenient. For the purpofe of merely removing a jar of air from one place to another, where it is not to ftand longer than a few days, I make ufe of common *tea-difhes*, which will hold water enough for that time, unlefs the air be in a ftate of diminution, by means of any procefs that is going on in it.

If I want to try whether an animal will live in any kind of air, I firft put the air into a fmall veffel, juft large enough to give it room to ftretch itfelf ; and as I generally make ufe of *mice* for this purpofe, I have found it very convenient to ufe the hollow part of a tall beer-glafs, *d.*fig. 1, which contains between two and
three

three ounce meafures of air. In this veffel a moufe will live twenty minutes, or half an hour.

For the purpofe of thefe experiments it is moft convenient to catch the mice in fmall wire traps, out of which it is eafy to take them, and holding them by the back of the neck, to pafs them through the water into the veffel which contains the air. If I expect that the moufe will live a confiderable time, I take care to put into the veffel fomething on which it may conveniently fit, out of the reach of the water. If the air be good, the moufe will foon be perfectly at its eafe, having fuffered nothing by its paffing through the water. If the air be fuppofed to be noxious, it will be proper (if the operator be defirous of preferving the mice for farther ufe) to keep hold of their tails, that they may be withdrawn as foon as they begin to fhew figns of uneafinefs; but if the air be thoroughly noxious, and the moufe happens to get a full infpiration, it will be impoffible to do this before it be abfolutely irrecoverable.

In order to *keep* the mice, I put them into receivers open at the top and bottom, ftanding upon plates of tin perforated with many holes, and covered with other plates of the fame kind, held down by fufficient weights, as fig. 3. Thefe
receivers

receivers ftand upon *a frame of wood,* that the frefh air may have an opportunity of getting to the bottoms of them, and circulating through them. In the infide I put a quantity of paper or tow, which muft be changed, and the veffel wafhed and dried, every two or three days. This is moft conveniently done by having another receiver, ready cleaned and prepared, into which the mice may be transferred till the other fhall be cleaned.

Mice muft be kept in a pretty exact temperature, for either much heat or much cold kills them prefently. The place in which I have generally kept them is a fhelf over the kitchen fire-place where, as it is ufual. in Yorkfhire, the fire never goes out; fo that the heat varies very little, and I find it to be, at a medium, about 70 degrees of Fahrenheit's thermometer. When they had been made to pafs through the water, as they neceffarily muft be in order to a change of air, they require, and will bear a very confiderable degree of heat, to warm and dry them.

I found, to my great furprize, in the courfe of thefe experiments, that mice will live intirely without water; for though I have kept them for three or four months, and have offered them water feveral times, they would never tafte it;
and

and yet they continued in perfect health and vigour. Two or three of them will live very
peaceably together in the fame veffel; though
I had one inftance of a moufe tearing another
almoft in pieces, and when there was plenty of
provifions for both of them.

In the fame manner in which a moufe is put
into a veffel of any kind of air, a *plant*, or any
thing elfe, may be put into it, viz. by paffing
it through the water; and if the plant be of a
kind that will grow in water only, there will be
no occafion to fet it in a pot of earth, which
will otherwife be neceffary.

There may appear, at firft fight, fome difficulty in opening the mouth of a phial, containing any fubftance, folid or liquid, to which
water muft not be admitted, in a jar of any
kind of air, which is an operation that I have
fometimes had recourfe to ; but this I eafily effect by means of *a cork cut tapering*, and a ftrong
wire thruft through it, as in fig. 4, for in this
form it will fufficiently fit the mouth of any
phial, and by holding the phial in one hand,
and the wire in the other, and plunging both
my hands into the trough of water, I can eafily
convey the phial through the water into the jar;
which muft either be held by an affiftant, or be
faftened by ftrings, with its mouth projecting

over

over the fhelf When the phial is thus conveyed into the jar, the cork may eafily be removed, and may alfo be put into it again at
pleafure, and conveyed the fame way out
again.

When any thing, as a gallipot, &c. is to be
fupported at a confiderable height within a jar,
it is convenient to have fuch *wire ftands* as are
reprefented fig. 5. They anfwer better than
any other, becaufe they take up but little room,
and may be eafily bended to any fhape or
height.

If I have occafion to pour air from a veffel
with a wide mouth into another with a very narrow one, I am obliged to make ufe of a *funnel*,
fig. 6, but by this means the operation is exceedingly eafy ; firft filling the veffel into which
the air is to be conveyed with water, and holding the mouth of it, together with the funnel,
both under water with one hand, while the
other is employed in pouring the air ; which,
afcending through the funnel up into the veffel,
makes the water defcend, and takes its place.
Thefe funnels are beft made of glafs, becaufe
the air being vifible through them, the quantity
of it may be more eafily eftimated by the eye.

It

It will be convenient to have feveral of thefe funnels of different fizes.

In order to expel air from folid fubftances by means of heat, I fometimes put them into a *gun-barrel*, fig. 7, and filling it up with dry fand, that has been well burned, fo that no air can come from it, I lute to the open end the ftem of a tobacco pipe, or a fmall glafs tube. Then having put the clofed end of the barrel, which contains the materials, into the fire, the generated air, iffuing through the tube, may be received in a veffel of quickfilver, with its mouth immerfed in a bafon of the fame, fufpended all together in wires, in the manner defcribed in the figure : or any other fluid fubftance may be ufed inftead of quickfilver.

But the moft accurate method of procuring air from feveral fubftances, by means of heat, is to put them, if they will bear it, into phials full of quickfilver, with the mouths immerfed in the fame, and then throw the focus of a burning mirror upon them. For this purpofe the phials fhould be made with their bottoms round, and very thin, that they may not be liable to break with a pretty fudden application of heat.

If

If I want to expel air from any liquid, I nearly fill a phial with it, and having a cork perforated, I put through it, and secure with cement, a glafs tube, bended in the manner reprefented at *e* fig. 1. I then put the phial into a kettle of water, which I fet upon the fire and make to boil. The air expelled by the heat, from the liquor contained in the phial, iffues through the tube, and is received in the bafon of quickfilver, fig. 7. Inftead of this fufpended bafon, I fometimes content myfelf with tying a flaccid bladder to the end of the tube, in both thefe proceffes, that it may receive the newly generated air

In experiments on thofe kinds of air which are readily imbibed by water, I always make ufe of quickfilver, in the manner reprefented fig. 8, in which *a* is the bafon of quickfilver, *b* a glafs veffel containing quickfilver, with its mouth immerfed in it, *c* a phial containing the ingredients from which the air is to be produced; and *d* is a fmall recipient, or glafs veffel defigned to receive and intercept any liquor that may be difcharged along with the air, which is to be tranfmitted free from any moifture into the veffel *b*. If there be no apprehenfion of moifture, I make ufe of the glafs tube only, without any recipient, in the manner reprefented *e* fig. 1. In order to invert the veffel *b*, I firft

fill

fill it with quickfilver, and then carefully cover the mouth of it with a piece of foft leather; after-which it may be turned upfide down without any danger of admitting the air, and the leather may be withdrawn when it is plunged in the quickfilver.

In order to generate air by the folution of metals, or any procefs of a fimilar nature, I put the materials into a phial, prepared in the manner reprefented at *e* fig. 1, and put the end of the glafs tube under the mouth of any veffel into which I want to convey the air. If heat be neceffary I can eafily apply it to a candle, or a red hot poker while it hangs in this pofition.

When I have occafion to transfer air from a jar ftanding in the trough of water to a veffel ftanding in quickfilver, or in any other fituation whatever, I make ufe of the contrivance reprefented fig. 9, which confifts of a bladder, fur-nifhed at one end with a fmall glafs tube bended, and at the other with a cork, perforated fo as juft to admit the fmall end of a funnel. When the common air is carefully preffed out of this bladder, and the funnel is thruft tightly into the cork, it may be filled with any kind of air as eafily as a glafs jar; and then a ftring being tied above the cork in which the funnel is in-
ferted,

ferted, and the orifice in the other cork clofed, by preffing the bladder againft it, it may be carried to any place, and if the tube be carefully wiped, the air may be conveyed quite free from moifture through a body of quickfilver, or any thing elfe. A little practice will make this very ufeful manœuvre perfectly eafy and accurate.

In order to impregnate fluids with any kind of air, as water with fixed air, I fill a phial with the fluid larger or lefs as I have occafion (as *a* fig. 10;) and then inverting it, place it with its mouth downwards, in a bowl *b*, containing a quantity of the fame fluid ; and having filled the bladder, fig. 9, with the air, I throw as much of it as I think proper into the phial, in the manner defcribed above. To accelerate the impregnation, I lay my hand on the top of the phial, and fhake it as much as I think proper.

If, without having any air previoufly generated, I would convey it into the fluid immediately as it arifes from the proper materials, I keep the fame bladder in connection with a phial *c* fig. 10, containing the fame materials (as chalk, falt of tartar, or pearl afhes in diluted oil of vitriol, for the generation of fixed air) and taking care (left, in the act of effervefcence, any of the materials in the phial *c* fhould

fhould get into the veffel *a*, to place this phial on a ftand lower than that on which the bafon was placed, I prefs out the newly generated air, and make it afcend directly into the fluid. For this purpofe, and that I may more conveniently fhake the phial *c*, which is neceffary in fome proceffes, efpecially with chalk and oil of vitriol, I fometimes make ufe of a flexible leathern tube *d*, and fometimes only a glafs tube. For if the bladder be of a fufficient length, it will give room for the agitation of the phial; or if not, it is eafy to connect two bladders together by means of a perforated cork, to which they may both be faftened.

When I want to try whether any kind of air will admit a candle to burn in it, I make ufe of a cylindrical glafs veffel, fig. 11. and a bit of wax candle *a* fig. 12, faftened to the end of a wire *b*, and turned up, in fuch a manner as to be let down into the veffel with the flame upwards. The veffel fhould be kept carefully covered till the moment that the candle is admitted. In this manner I have frequently extinguifhed a candle more than twenty times fucceffively, in a veffel of this kind, though it is impoffible to dip the candle into it without giving the external air an opportunity of mixing with the air in the infide more or lefs. The candle *c*, at the other end of the wire is

C   very

very convenient for holding under a jar ſtanding in water, in order to burn as long as the incloſed air can ſupply it; for the moment that it is extinguiſhed, it may be drawn through the water before any ſmoke can have mixed with the air.

In order to draw air out of a veſſel which has its mouth immerſed in water, and thereby to raiſe the water to whatever height may be neceſſary, it is very convenient to make uſe of a glaſs *ſyphon*, fig. 13, putting one of the legs up into the veſſel, and drawing the air out at the other end by the mouth. If the air be of a noxious quality, it may be neceſſary to have a ſyringe faſtened to the ſyphon, the manner of which needs no explanation. I have not thought it ſafe to depend upon a valve at the top of the veſſel, which Dr. Hales ſometimes made uſe of.

If, however, a very ſmall hole be made at the top of a glaſs veſſel, it may be filled to any height by holding it under water, while the air is iſſuing out at the hole, which may then be cloſed with wax or cement.

If the generated air will neither be abſorbed by water, nor diminiſh common air, it may be convenient to put part of the materials into a cup, ſupported by a ſtand, and the other part into

into a fmall glafs veffel, placed on the edge of
it, as at *f*, fig. 1. Then having, by means of a
fyphon, drawn the air to a convenient height,
the fmall glafs veffel may be eafily pufhed into
the cup, by a wire introduced through the water ; or it may be contrived, in a variety of
ways, only to difcharge the contents of the
fmall veffel into the larger. The diftance between the boundary of air and water, before and
after the operation, will fhew the quantity of
the generated air. The effect of proceffes that
diminifh air may alfo be tried by the fame apparatus.

When I want to admit a particular kind
of air to any thing that will not bear wetting,
and yet cannot be conveniently put into a phial,
and efpecially if it be in the form of a powder,
and muft be placed upon a ftand (as in thofe
experiments in which the focus of a burning
mirror is to be thrown upon it) I firft exhauft
a receiver, in which it is previoufly placed ;
and having a glafs tube, bended for the purpofe, as in fig. 14, I fcrew it to the ftem of a
transfer of the air pump on which the receiver
had been exhaufted, and introducing it through
the water into a jar of that kind of air with
which I would fill the receiver, I only turn
the cock, and I gain my purpofe. In this
method, however, unlefs the pump be very

good,

good, and feveral contrivances, too minute to
be particularly defcribed, be made ufe of a
good deal of common air will get into the receiver.

When I want to meafure the goodnefs of
any kind of air, I put two meafures of it into
a jar ftanding in water ; and when I have
marked upon the glafs the exact place of the
boundary of air and water, I put to it one
meafure of nitrous air : and after waiting a
proper time, note the quantity of its diminution. If I be comparing two kinds of air
that are nearly alike, after mixing them in a
large jar, I transfer the mixture into a long
glafs tube, by which I can lengthen my fcale
to what degree I pleafe.

If the quantity of the air, the goodnefs of
which I want to afcertain, be exceedingly
fmall, fo as to be contained in a part of a glafs
tube, out of which water will not run fpontaneoufly, as a fig. 15; I firft meafure with a
pair of compaffes the length of the column
of air in the tube, the remaining part being
filled with water, and lay it down upon a
fcale ; and then, thrufting a wire of a proper
thicknefs, b, into the tube, I contrive, by
means of a thin plate of iron, bent to a fharp
angle c, to draw it out again, when the whole

of

of this little apparatus has been introduced through the water into a jar of nitrous air; and the wire being drawn out, the air from the jar muſt ſupply its place. I then meaſure the length of this column of nitrous air which I have got into the tube, and lay it alſo down upon the ſcale, ſo as to know the exact length of both the columns. After this, holding the tube under water, with a ſmall wire I force the two ſeparate columns of air into contact, and when they have been a ſufficient time together, I meaſure the length of the whole, and compare it with the length of both the columns taken before. A little experience will teach the operator how far to thruſt the wire into the tube, in order to admit as much air as he wants and no more.

In order to take the electric ſpark in a quantity of any kind of air, which muſt be very ſmall, to produce a ſenſible effect upon it, in a ſhort time, by means of a common machine, I put a piece of wire into the end of a ſmall tube, and faſten it with hot cement, as in fig. 16; and having got the air I want into the tube by means of the apparatus fig. 15, I place it inverted in a baſon containing either quickſilver, or any other fluid ſubſtance by which I chuſe to have the air confined. I then, by the help of the air pump, drive out as much of the air as I think convenient, admitting the

C 3   quick-

quickfilver, &c. to it, as at *a*, and putting a brafs ball on the end of the wire, I take the fparks or fhocks upon it, and thereby tranfmit them through the air to the liquor in the tube.

To take the electric fparks in any kind of fluid, as oil, &c. I ufe the fame apparatus defcribed above, and having poured into the tube as much of the fluid as I conjecture I can make the electric fpark pafs through, I fill the reft with quickfilver; and placing it inverted in a bafon of quickfilver, I take the fparks as before.

If air be generated very faft by this procefs, I ufe a tube that is narrow at the top, and grows wider below, as fig. 17, that the quickfilver may not recede too foon beyond the ftriking diftance.

Sometimes I have ufed a different apparatus for this purpofe, reprefented fig. 18. Taking a pretty wide glafs tube, hermetically fealed at the upper-end, and open below, at about an inch, or at what diftance I think convenient from the top, I get two holes made in it, oppofite to each other. Through thefe I put two wires, and faftening them with warm cement, I fix them at what diftance I pleafe from each other. Between thefe wires I take the fparks, and the bubbles of air rife, as they are formed, to the top of the tube.

PART

PART I.

*Experiments and Observations made in, and be-
fore the year* 1772.

IN writing upon the fubject of *different kinds of
air,* I find myfelf at a lofs for proper *terms,*
by which to diftinguifh them, thofe which
have hitherto obtained being by no means fufficiently characteriftic, or diftinct. The only terms
in common ufe are, *fixed air, mephitic,* and *inflammable.* The laft, indeed, fufficiently cha-
racterizes and diftinguifhes that kind of air
which takes fire, and explodes on the approach
of flame; but it might have been termed *fixed*
with as much propriety as that to which Dr.
Black and others have given that denomination,
fince it is originally part of fome folid fubftance,
and exifts in an unelaftic ftate.

All

All thefe newly difcovered kinds of air
may alfo be called *factitious*; and if, with
others, we ufe the term *fixable*, it is ftill obvious
to remark, that it is applicable to them all;
fince they are all capable of being imbibed by
fome fubftance or other, and confequently of
being *fixed* in them, after they have been in an
elaftic ftate.

The term *mephitic* is equally applicable
to what is called *fixed air*, to that which is *inflammable*, and to many other kinds; fince they
are equally noxious, when breathed by animals.
Rather, however, than either introduce new
terms, or change the fignification of old ones,
I fhall ufe the term *fixed air*, in the fenfe in
which it is now commonly ufed, and diftinguifh
the other kinds by their properties, or fome
other periphrafis. I fhall be under a neceffity,
however, of giving names to thofe kinds of
air, to which no names has been given by o-
thers, as *nitrous*, *acid*, and *alkaline*.

SECT.

S E C T I O N I.

Of FIXED AIR.

It was in confequence of living for fome time in the neighbourhood of a public brewery, that I was induced to make experiments on fixed air, of which there is always a large body, ready formed, upon the furface of the fermenting liquor, generally about nine inches, or a foot in depth, within which any kind of fubftance may be very conveniently placed ; and though, in thefe circumftances, the fixed air muft be continually mixing with the common air, and is therefore far from being perfectly pure, yet there is a conftant frefh fupply from the fermenting liquor, and it is pure enough for many purpofes.

A perfon, who is quite a ftranger to the properties of this kind of air, would be agreeably amufed with extinguifhing lighted candles, or chips of wood in it, as it lies upon the furface of the fermenting liquor ; for the fmoke readily unites with this kind of air, probably by means of the water which it contains ; fo that very little or none of the fmoke will efcape into the open air, which is incumbent upon it. It is remarkable, that the upper furface of this

fmoke

ſmoke, floating in the fixed air, is ſmooth, and
well defined ; whereas the lower ſurface is exceedingly ragged, ſeveral parts hanging down to
a conſiderable diſtance within the body of the
fixed air, and ſometimes in the form of balls,
connected to the upper ſtratum by ſlender
threads, as if they were ſuſpended. The ſmoke
is alſo apt to form itſelf into broad flakes, pa-
rallel to the ſurface of the liquor, and at different diſtances from it, exactly like clouds.
Theſe appearances will ſometimes continue a-
bove an hour, with very little variation. When
this fixed air is very ſtrong, the ſmoke of a
ſmall quantity of gunpowder fired in it will be
wholly retained by it, no part eſcaping into the
common air,

Making an agitation in this air, the ſurface
of it, (which ſtill continues to be exactly defined) is thrown into the form of waves, which
it is very amuſing to look upon ; and if, by
this agitation, any of the fixed air be thrown
over the ſide of the veſſel, the ſmoke, which
is mixed with it, will fall to the ground, as if
it was ſo much water, the fixed air being
heavier than common air.

The red part of burning wood was extinguiſhed in this air, but I could not perceive that
a red-hot poker was ſooner cooled in it.

Fixed

Fixed air does not inftantly mix with common air. Indeed if it did, it could not be caught upon the furface of the fermenting liquor. A candle put under a large receiver, and immediately plunged very deep below the furface of the fixed air, will burn fome time. But veffels with the fmalleft orifices, hanging with their mouths downwards in the fixed air, will *in time* have the common air, which they contain, perfectly mixed with it. When the fermenting liquor is contained in veffels clofe covered up, the fixed air, on removing the cover, readily affects the common air which is contiguous to it; fo that, candles held at a confiderable diftance above the furface will inftantly go out. I have been told by the workmen, that this will fometimes be the cafe, when the candles are held two feet above the mouth of the veffel.

Fixed air unites with the fmoke of rofin, fulphur, and other electrical fubftances, as well as with the vapour of water; and yet, by holding the wire of a charged phial among thefe fumes, I could not make any electrical atmofphere, which furprized me a good deal, as there was a large body of this fmoke, and it was fo confined, that it could not efcape me.

I alfo held fome oil of vitriol in a glafs veffel within the fixed air, and by plunging a
piece

piece of red-hot glafs into it, raifed a copious and thick fume. This floated upon the furface of the fixed air like other fumes, and continued as long.

Confidering the near affinity between water and fixed air, I concluded that if a quantity of water was placed near the yeaft of the fermenting liquor, it could not fail to imbibe that air, and thereby acquire the principal properties of Pyrmont, and fome other medicinal mineral waters. Accordingly, I found, that when the furface of the water was confiderable, it always acquired the pleafant acidulous tafte that Pyrmont water has. The readieft way of impregnating water with this virtue, in thefe circumftances, is to take two veffels, and to keep pouring the water from one into the other, when they are both of them held as near the yeaft as poffible ; for by this means a great quantity of furface is expofed to the air, and the furface is alfo continually changing. In this manner, I have fometimes, in the fpace of two or three minutes, made a glafs of exceedingly pleafant fparkling water, which could hardly be diftinguifhed from very good Pyrmont, or rather Seltzer water.

But the *moft effectual* way of impregnating water with fixed air is to put the veffels which contain the water into glafs jars, filled with the
pureft

pureft fixed air made by the folution of chalk
in diluted oil of vitriol, ftanding in quickfilver.
In this manner I have, in about two days, made
a quantity of water to imbibe more than an
equal bulk of fixed air, fo that, according to
Dr. Brownrigg's experiments, it muft have been
much ftronger than the beft imported Pyrmont;
for though he made his experiments at the
fpring-head, he never found that it contained
quite fo much as half its bulk of this air. If a
fufficient quantity of quickfilver cannot be procured, *oil* may be ufed with fufficient advantage,
for this purpofe, as it imbibes the fixed air very
flowly.   Fixed air may be kept in veffels ftanding in water for a long time, if they be feparated by a partition of oil, about half an inch
thick.   Pyrmont water made in thefe circumftances, is little or nothing inferior to that
which has ftood in quickfilver.

The *readieft* method of preparing this water
for ufe is to agitate it ftrongly with a large
furface expofed to the fixed air. By this means
more than an equal bulk of air may be communicated to a large quantity of water in the
fpace of a few minutes.   But fince agitation
promotes the diffipation of fixed air from water, it cannot be made to imbibe fo great a
quantity in this method as in the former,
where more time is taken.

Eafy

Eafy directions for impregnating water with fixed air I have publifhed in a fmall pamphlet, defigned originally for the ufe of feamen in long voyages, on the prefumption that it might be of ufe for preventing or curing the fea fcurvy, equally with wort, which was recommended by Dr. Macbride for this purpofe, on no other account than its property of generating fixed air, by its fermentation in the ftomach.

Water thus impregnated with fixed air readily diffolves iron, as Mr. Lane has difcovered; fo that if a quantity of iron filings be put to it, it prefently becomes a ftrong chalybeate, and of the mildeft and moft agreeable kind.

I have recommended the ufe of *chalk* and oil of vitriol as the cheapeft, and, upon the whole, the beft materials for this purpofe. But fome perfons prefer *pearl afhes*, *pounded marble*, or other calcareous or *alkaline fubftances*, and perhaps with reafon. My own experience has not been fufficient to enable me to decide in this cafe.

Whereas fome perfons had fufpected that a quantity of the oil of vitriol was rendered volatile by this procefs, I examined it, by all the chemical methods that are in ufe; but could

not

not find that water thus impregnated contained the leaft perceivable quantity of that acid.

Mr. Hey, indeed, who affifted me in this examination, found that diftilled water, impregnated with fixed air, did not mix fo readily with foap as the diftilled water itfelf; but this was alfo the cafe when the fixed air had paffed through a long glafs tube filled with alkaline falts, which, it may be fuppofed, would have imbibed any of the oil of vitriol that might have been contained in that air *.

Fixed air itfelf may be faid to be of the nature of an acid, though of a weak and peculiar fort.——Mr. Bergman of Upfal, who honoured me with a letter upon the fubject, calls it the *aërial acid*, and, among other experiments to prove it to be an acid, he fays that it changes the blue juice of tournefole into red. This Mr. Hey found to be true, and he moreover difcovered that when water tinged blue with the juice of tournefole, and then red with fixed air, has been expofed to the open air, it recovers its blue colour again.

The heat of boiling water will expel all the fixed air, if a phial containing the impregnated

* An account of Mr. Hey's experiments will be found in the Appendix to thefe papers.

water

water be held in it ; but it will often require
above half an hour to do it completely.

Dr. Percival, who is particularly attentive
to every improvement in the medical art, and
who has thought fo well of this impregnation
as to prefcribe it in feveral cafes, informs me
that it feems to be much ftronger, and fparkles
more, like the true Pyrmont water, after it
has been kept fome time. This circumftance,
however, fhews that, in time, the fixed air is
more eafily difengaged from the water ; and
though, in this ftate, it may affect the tafte
more fenfibly, it cannot be of fo much ufe
in the ftomach and bowels, as when the air is
more firmly retained by the water.

By the procefs defcribed in my pamphlet,
fixed air may be readily incorporated with wine,
beer, and almoft any other liquor whatever ;
and when beer, wine, or cyder, is become
flat or dead (which is the confequence of the
efcape of the fixed air they contained) they
may be revived by this means ; but the delicate and agreeable flavour, or acidulous tafte,
communicated by fixed air, and which is very
manifeft in water, can hardly be perceived in
wine, or any liquors which have much tafte
of their own.

I fhould

I fhould think that there can be no doubt, but that water thus impregnated with fixed air muft nave all the medicinal virtues of genuine Pyrmont or Seltzer water; fince thefe depend upon the fixed air they contain. If the genuine Pyrmont water derives any advantage from its being a natural chalybeate, this may alfo be obtained by providing a common chalybeate water, and ufing it in thefe procefles, inftead of common water.

Having fucceeded fo well with this artificial Pyrmont water, I imagined that it might be poffible to give *ice* the fame virtue, efpecially as cold is known to promote the abforption of fixed air by water ; but in this I found myfelf quite miftaken. I put feveral pieces of ice into a quantity of fixed air, confined by quickfilver, but no part of the air was abforbed in two days and two nights ; but upon bringing it into a place where the ice melted, the air was abforbed as ufual.

I then took a quantity of ftrong artificial Pyrmont water, and putting it into a thin glafs phial, l fet it in a pot that was filled with fnow and falt. This mixture inftantly freezing the water that was contiguous to the fides of the glafs, the air was difcharged plentifully, fo

D   that

that I catched a conſiderable quantity, in a
bladder tied to the mouth of the phial.

I alſo took two quantities of the ſame Pyrmont water, and placed one of them where it
might freeze, keeping the other in a cold place,
but where it would not freeze. This retained
its acidulous taſte, though the phial which
contained it was not corked ; whereas the other
being brought into the ſame place, where the
ice melted very ſlowly, had at the ſame time
the taſte of common water only. That quantity of water which had been frozen by the
mixture of ſnow and ſalt, was almoſt as much
like ſnow as ice, ſuch a quantity of air-bubbles were contained in it, by which it was prodigiouſly increaſed in bulk.

The preſſure of the atmoſphere aſſiſts very
conſiderably in keeping fixed air confined in
water ; for in an exhauſted receiver, Pyrmont
water will abſolutely boil, by the copious diſcharge of its air. This is alſo the reaſon why
beer and ale froth ſo much *in vacuo.* I do not
doubt, therefore, but that, by the help of a
condenſing engine, water might be much more
highly impregnated with the virtues of the
Pyrmont ſpring ; and it would not be difficult
to contrive a method of doing it.

The

The manner in which I made feveral experiments to afcertain the abforption of fixed air
by different fluid fubftances, was to put the liquid into a difh, and holding it within the
body of the fixed air at the brewery, to fet a
glafs veffel into it, with its mouth inverted.
This glafs being neceffarily filled with the fixed
air, the liquor would rife into it when they
were both taken into the common air, if the
fixed air was abforbed at all.

Making ufe of *ether* in this manner, there
was a conftant bubbling from under the glafs,
occafioned by this fluid eafily rifing in vapour,
fo that I could not, in this method, determine
whether it imbibed the air or not. I concluded
however, that they did incorporate, from a
very difagreeable circumftance, which made me
defift from making any more experiments of
the kind. For all the beer, over which this
experiment was made, contracted a peculiar
tafte ; the fixed air impregnated with the ether
being, I fuppofe, again abforbed by the beer.
I have alfo obferved, that water which remained a long time within this air has fometimes
acquired a very difagreeable tafte. At one
time it was like tar-water. How this was acquired, I was very defirous of making fome
experiments to afcertain, but I was difcouraged

by the fear of injuring the fermenting liquor. It could not come from the fixed air only.

Infects and animals which breathe very little are ſtifled in fixed air, but are not ſoon quite killed in it. Butterflies and flies of other kinds will generally become torpid, and ſeem-ingly dead, after being held a few minutes over the fermenting liquor ; but they revive again after being brought into the freſh air. But there are very great varieties with refpe&t to the time in which different kinds of flies will either become torpid in the fixed air, or die in it. A large ſtrong frog was much ſwelled, and ſeemed to be nearly dead, after being held about fix minutes over the fermenting liquor ; but it recovered upon being brought into the common air. A ſnail treated in the ſame man-ner died prefently.

Fixed air is prefently fatal to vegetable life. At leaſt ſprigs of mint growing in water, and placed over the fermenting liquor, will often become quite dead in one day, or even in a Jeſs ſpace of time ; nor do they recover when they are afterwards brought into the common aır. I am told, however, that ſome other plants are much more hardy in this refpeČt.

A red

A red rofe, frefh gathered, loft its rednefs, and became of a purple colour, after being held over the fermenting liquor about twenty-four hours ; but the tips of each leaf were much more affected than the reft of it. Another red rofe turned perfectly white in this fituation ; but various other flowers of different colours were very little affected. Thefe experiments were not repeated, as I wifh they might be done, in pure fixed air, extracted from chalk by means of oil of vitriol.

For every purpofe, in which it was neceffary that the fixed air fhould be as unmixed as poffible, I generally made it by pouring oil of vitriol upon chalk and water, catching it in a bladder faftened to the neck of the phial in which they were contained, taking care to prefs out all the common air, and alfo the firft, and fometimes the fecond, produce of fixed air ; and alfo, by agitation, making it as quickly as I poffibly could. At other times, I made it pafs from the phial in which it was generated through a glafs tube, without the intervention of any bladder, which, as I found by experience, will not long make a fufficient feparation between feveral kinds of air and common air.

I had once thought that the readieft method of procuring fixed air, and in fufficient purity,

would

would be by the fimple procefs of burning
chalk, or pounded lime-ftone in a gun-barrel,
making it pafs through the ftem of a tobacco-
pipe, or a glafs tube carefully luted to the
orifice of it.   In this manner I found that air
is produced in great plenty; but, upon examining it, I found, to my very great furprife,
that little more than one half of it was fixed
air, capable of being abforbed by water ; and
that the reft was inflammable, fometimes very
weakly, but fometimes pretty highly fo.

Whence this inflammability proceeds, I am
not able to determine, the lime or chalk not
being fuppofed to contain any other than fixed
air.   I conjecture, however, that it muft proceed from the iron, and the feparation of it
from the calx may be promoted by that fmall
quantity of oil of vitriol, which I am informed
is contained in chalk, if not in lime-ftone alfo.

But it is an objection to this hypothefis, that
the inflammable air produced in this manner
burns blue, and not at all like that which is produced from iron, or any other metal, by means
of an acid.   It alfo has not the fmell of that kind
of inflammable air which is produced from mineral fubftances.   Befides, oil of vitriol without water, will not diffolve iron ; nor can inflammable air be got from it, unlefs the acid be
con-

confiderably diluted ; and when I mixed brimftone with the chalk, neither the quality nor
the quantity of the air was changed by it. Indeed no air, or permanently elaftic vapour, can
be got from brimftone, or any oil.

Perhaps this inflammable principle may
come from fome remains of the animals, from
which it is thought that all calcareous matter
proceeds.

In the method in which I generally made the
fixed air (and indeed always, unlefs the contrary be particularly mentioned, *viz.* by diluted
oil of vitriol and chalk) I found by experiment
that it was as pure as Mr. Cavendifh made it.
For after it had paffed through a large body of
water in fmall bubbles, ftill $\frac{1}{30}$ or $\frac{1}{60}$ part only
was not abforbed by water. In order to try
this as expeditioufly as poffible, I kept pouring
the air from one glafs veffel into another, immerfed in a quantity of cold water, in which
manner I found by experience, that almoft any
quantity may be reduced as far as poffible in a
very fhort time. But the moft expeditious method of making water imbibe any kind of air,
is to confine it in a jar; and agitate it ftrongly, in
the manner defcribed in my pamphlet on the
impregnation of water with fixed air, and reprefented fig. 10.

D 4

At

At the ſame time that I was trying the purity
of my fixed air, I had the curioſity to endeavour
to aſcertain whether that part of it which is not
miſcible in water, be equally diffuſed through
the whole maſs ; and, for this purpoſe, I divided a quantity of about a gallon into three
parts, the firſt conſiſting of that which was up-
permoſt, and the laſt of that which was the
loweſt, contiguous to the water ; but all theſe
parts were reduced in about an equal proportion, by paſſing through the water, ſo that the
whole maſs had been of an uniform compoſition. This I have alſo found to be the caſe
with ſeveral kinds of air, which will not properly incorporate.

A mouſe will live very well, though a candle will not burn in the reſiduum of the pureſt
fixed air that I can make ; and I once made a
very large quantity for the ſole purpoſe of this
experiment. This, therefore, ſeems to be one
inſtance of the generation of genuine common
air, though vitiated in ſome degree. It is alſo
another proof of the reſiduum of fixed air being,
in part at leaſt, common air, that it becomes
turbid, and is diminiſhed by the mixture of nitrous air, as will be explained hereafter.

That fixed air only wants ſome addition to
make it permanent, and immiſcible with water
I   if

if not in all refpects, common air, 1 have been
led to conclude, from feveral attepmts which I
once made to mix it with air in which a quantity of iron filings and brimftone, made into a
pafte witn water, had ftood; for, in feveral
mixtures of this kind, I imagined that not
much more than half of the fixed air could be
imbibed by water; but, not being able to repeat the experiment, I conclude that I either
deceived myfelf in it, or that I overlooked
fome circumftance on which the fuccefs of it
depended.

Thefe experiments, however, whether they
were fallacious or otherwife, induced me to try
whether any alteration would be made in the
conftitution of fixed air, by this mixture of
iron filings and brimftone. I therefore put a
mixture cf this kind into a quantity of as pure
fixed air as I could make, and confined the
whole in quickfilver, left the water fhould abforb it before the effects of the mixture could
take place. The confequence was, that the
fixed air was diminifhed, and the quickfilver
rofe in the veffel, till about the fifth part was
occupied by it; and, as near as I could judge,
the procefs went on, in all refpects, as if the
air in the infide had been common air.

What

What is moſt remarkable, in the reſult of this experiment, is, that the fixed air, into which this mixture had been put, and which had been in part diminiſhed by it, was in part alſo rendered inſoluble in water by this means. I made this experiment four times, with the greateſt care, and oberved, that in two of them about one ſixth, and in the other two about one fourteenth, of the original quantity, was ſuch as could not be abſorbed by water, but continued permanently elaſtic. Left I ſhould have made any miſtake with reſpect to the purity of the fixed air, the laſt time that I made the experiment, I ſet part of the fixed air, which I made uſe of, in a ſeparate veſſel, and found it to be exceedingly pure, ſo as to be almoſt wholly abſorbed by water ; whereas the other part, to which I had put the mixture, was far from being ſo.

In one of theſe caſes, in which fixed air was made immiſcible with water, it appeared to be not very noxious to animals ; but in another caſe, a mouſe died in it pretty ſoon. This difference probably aroſe from my having inadvertently agitated the air in water rather more in one caſe than in the other.

As the iron is reduced to a calx by this proceſs, I once concluded, that it is phlogiſton that
                                                    fixed

fixed air wants, to make it common air; and, for any thing I yet know, this may be the cafe, though I am ignorant of the method of combining them; and when I calcined a quantity of lead in fixed air, in the manner which will be defcribed hereafter, it did not feem to have been lefs foluble in water than it was before.

SECTION II.

Of AIR *in which a* CANDLE, *or* BRIMSTONE, *has burned out.*

It is well known that flame cannot fubfift long without change of air, fo that the common air is neceffary to it, except in the cafe of fubftances, into the compofition of which nitre enters, for thefe will burn *in vacuo*, in fixed air, and even under water, as is evident in fome rockets, which are made for this purpofe. The quantity of air which even a fmall flame requires to keep it burning is prodigious. It is generally faid, that an ordinary candle *confumes*, as it is called, about a gallon in a minute. Confidering this amazing confumption of air, by fires of all kinds, volcanos, &c. it becomes a great objeft of philofophical inquiry, to afcertain what change is made in the conftitution of the air by flame, and to difcover what provifion

fion there is in nature for remedying the injury which the atmofphere receives by this means. Some of the following experiments will, perhaps, be thought to throw light upon the fubjeçt.

The diminution of the quantity of air in which a candle, or brimftone, has burned out, is various ; But I imagine that, at a medium, it may be about one fifteenth, or one fixteenth of the whole ; which is one third as much as by animal or vegetable fubftances putrefying in it, by the calcination of metals, or by any of the other caufes of the complete diminution of air, which will be mentioned hereafter.

I have fometimes thought, that flame difpofes the common air to depofit the fixed air it contains; for if any lime-water be expofed to it, it immediately becomes turbid. This is the cafe, when wax candles, tallow candles, chips of wood, fpirit of wine, ether, and every other fubftance which I have yet tried, except brimftone, is burned in a clofe glafs veffel, ftanding in lime-water. This precipitation of fixed air (if this be the cafe) may be owing to fomething emitted from the burning bodies, which has a ftronger affinity with the other conftituent parts of the atmofphere *.

* The fuppofition, mentioned in this and other paffages of the firft part of this publication, viz. that the diminution

of

If brimftone be burned in the fame circumftances, the lime water continues tranfparent,
but ftill there may have been the fame precipitation of the fixed part of the air ; but that,
uniting with the lime and the vitriolic acid, it
forms a felenetic falt, which is foluble in water.
Having evaporated a quantity of water thus
impregnated, by burning brimftone a great
number of times over it, a whitifh powder remained, which had an acid tafte; but repeating
the experiment with a quicker evaporation, the
powder had no acidity, but was very much like
chalk. The burning of brimftone but once
over a quantity of lime-water, will affect it in
fuch a manner, that breathing into it will not
make it turbid, which otherwife it always prefently does.

Dr. Hales fuppofed, that by burning brimftone repeatedly in the fame quantity of air, the
diminution would continue without end. But
this I have frequently tried, and not found to
be the cafe. Indeed, when the ignition has
been imperfect in the firft inftance, a fecond
firing of the fame fubftance will increafe the
effect of the firft, &c. but this progrefs foon
ceafes.

of common air, by this and other procefles is, in part at
leaft, owing to the precipitation of the fixed air from it,
the reader will find confirmed by the experiments and obfervations in the fecond part,

In

In many cafes of the diminution of air, the effect is not immediately apparent, even when it ftands in water ; for fometimes the bulk of air will not be much reduced, till it has paffed feveral times through a quantity of water, which has thereby a better opportunity of abforbing that part of the air, which had not been perfectly detatched from the reft. I have fometimes found a very great reduction of a mafs of air, in confequence of paffing but once through cold water. If the air has ftood in quickfilver, the diminution is generally inconfiderable, till it has undergone this operation, there not being any fubftance expofed to the air that could abforb any part of it.

I could not find any confiderable alteration in the fpecific gravity of the air, in which candles, or brimftone, had burned out. I am fatisfied, however, that it is not heavier than common air, which muft have been manifeft, if fo great a diminution of the quantity had been owing, as Dr. Hales and others fuppofed, to the elafticity of the whole mafs being impaired. After making feveral trials for this purpofe, I concluded that air, thus diminifhed in bulk, is rather lighter than common air, which favours the fuppofition of the fixed, or heavier part of the common air, having been precipitated.

An

An animal will live nearly, if not quite as long, in air in which candles have burned out, as in common air. This fact furprized me very greatly, having imagined that what is called the *confumption* of air by flame, or refpiration, to have been of the fame nature, and in the fame degree; but I have fince found, that this fact has been obferved by many perfons, and even fo early as by Mr. Boyle. I have alfo obferved, that air, in which brimftone has burned, is not in the leaft injurious to animals, after the fumes, which at firft make it very cloudy, have intirely fubfided.

I muft, in this place, admonifh my reader not to confound the fimple *burning of brimftone*, or of matches (*i. e.* bits of wood dipped in it) and the burning of brimftone with a burning mirror, or any *foreign heat*. The effect of the former is nothing more than that of any other *flame*, or *ignited vapour*, which will not burn, unlefs the air with which it is furrounded be in a very pure ftate, and which is therefore extinguifhed when the air begins to be much vitiated. Lighted brimftone, therefore reduces the air to the fame ftate as lighted wood. But the focus of a burning mirror thrown for a fufficient time either upon brimftone, or wood, after it has ceafed to burn of its own accord, and has become *charcoal*, will have a much

greater

greater effect of the fame kind, diminifhing the
air to its utmoft extent, and making it thoroughly noxious. In fact, as will be feen hereafter,
more phlogifton is expelled from thefe fubftances in the latter cafe than in the former.
I never, indeed, actually carried this experiment fo far with brimftone; but from the
diminution of air that I did produce by this
means, I concluded that, by continuing the
procefs fome time longer, it would have been
effected.

Having read, in the Memoirs of the Philofophical Society at Turin, vol. I. p. 41. that air in
which candles had burned out was perfectly
reftored, fo that other candles would burn in it
again as well as ever, after having been expofed
to a confiderable degree of *cold*, and likewife
after having been compreffed in bladders (for
the cold had been fuppofed to have produced
this effect by nothing but *condenfation*) I repeated thofe experiments, and did, indeed, find,
that when I compreffed the air in *bladders*, as
the Count de Saluce, who made the obfervation,
had done, the experiment fucceeded: but having
had fufficient reafon to diftruft bladders, I compreffed the air in a glafs veffel ftanding in water;
and then I found, that this procefs is altogether
ineffectual for the purpofe. I kept the air compreffed much more, and much longer, than the
                                                    Count

Count had done, but without producing any
alteration in it. I alfo find, that a greater degree of cold than that which he applied, and
of longer continuance, did by no means reftore
this kind of air: for when I had expofed the
phials which contained it a whole night, in
which the froft was very intenfe; and alfo when
I kept it furrounded with a mixture of fnow
and falt, I found it, in all refpects, the fame
as before.

It is alfo advanced, in the fame Memoir,
p. 41. that *heat* only, as the reverfe of *cold*,
renders air unfit for candles burning in it. But
I repeated the experiment of the Count for that
purpofe, without finding any fuch effect from
it. I alfo remember that, many years ago, I
filled an exhaufted receiver with air, which had
paffed through a glafs tube made red-hot, and
found that a candle would burn in it perfectly
well. Alfo, rarefaction by the air-pump does
not injure air in the leaft degree.

Though this experiment failed, I have been
fo happy, as by accident to have hit upon a
method of reftoring air, which has been injured
by the burning of candles, and to have difcovered at leaft one of the reftoratives which nature employs for this purpofe. It is *vegetation.*
This reftoration of vitiated air, I conjecture, is

E   effected

effected by plants imbibing the phlogiftic matter with which it is overloaded by the burning of inflammable bodies. But whether there be any foundation for this conjecture or not, the fact is, I think, indifputable. I fhall introduce the account of my experiments on this fubject, by reciting fome of the obfervations which I made on the growing of plants in confined air, which led to this difcovery.

One might have imagined that, fince common air is neceffary to vegetable, as well as to animal life, both plants and animals had affected it in the fame manner ; and I own I had that expectation, when I firft put a fprig of mint into a glafs jar, ftanding inverted in a veffel of water : but when it had continued growing there for fome months, I found that the air would neither extinguifh a candle, nor was it at all inconvenient to a moufe, which I put into it.

The plant was not affected any otherwife than was the neceffary confequence of its confined fituation ; for plants growing in feveral other kinds of air, were all affected in the very fame manner. Every fucceffion of leaves was more diminifhed in fize than the preceding, till, at length, they came to be no bigger than the heads of pretty fmall pins. The root decayed, and the ftalk alfo, beginning from the root ;
and

and yet the plant continued to grow upwards,
drawing its nourifhment through a black and
rotten ftem. In the third or fourth fet of
leaves, long and white hairy filaments grew
from the infertion of each leaf and fometimes
from the body of the ftem, fhooting out as far
as the veffel in which it grew would permit,
which, in my experiments, was about two inches. In this manner a fprig of mint lived
the old plant decaying, and new ones fhooting
up in its place, but lefs and lefs continually,
all the fummer feafon.

In repeating this experiment, care muft be
taken to draw away all the dead leaves from
about the plant, left they fhould putrefy, and
affect the air. I have found that a frefh cabbage leaf, put under a glafs veffel filled with
common air, for the fpace of one night only,
has fo affected the air, that a candle would not
burn in it the next morning, and yet the leaf
had not acquired any fmell of putrefaction.

Finding that candles would burn very well
in air in which plants had grown a long time,
and having had fome reafon to think, that there
was fomething attending vegetation, which reftored air that had been injured by refpiration,
I thought it was poffible that the fame procefs

E 2   might

might alfo reftore the air that had been
injured by the burning of candles.

Accordingly, on the 17th of Auguft 1771,
I put a fprig of mint into a quantity of air, in
which a wax candle had burned out, and found
that, on the 27th of the fame month, another
candle burned perfectly well in it. This experiment I repeated, without the leaft variation
in the event, not lefs than eight or ten times in
the remainder of the fummer.

Several times I divided the quantity of air in
which the cardle had burned out, into two
parts, and putting the plant into one of them,
left the other in the fame expofure, contained,
alfo, in a glafs veffel immerfed in water, but
without any plant; and never failed to find,
that a candle would burn in the former, but
not in the latter.

I generally found that five or fix days were
fufficient to reftore this air, when the plant was
in its vigour; whereas I have kept this kind of
air in glafs veffels, immerfed in water many
months, without being able to perceive that the
leaft alteration had been made in it. I have
alfo tried a great variety of experiments upon
it, as by condenfing, rarefying, expofing to the
light and heat, &c. and throwing into it the
                                        effluvia

effluvia of many different fubftances, but without any effect.

Experiments made in the year 1772, abun-
dantly confirmed my conclufion concerning the
reftoration of air, in which candles had burned
out by plants growing in it. The firft of thefe
experiments was made in the month of May ;
and they were frequently repeated in that and
the two following months, without a fingle
failure.

For this purpofe I ufed the flames of different
fubftances, though I generally ufed wax or tallow candles. On the 24th of June the experiment fucceeded perfectly well with air in
which fpirit of wine had burned out, and on
the 27tn of the famen onth it futceeded equally
well with air in which brimftone matches had
burned out, an effect of which I had defpaired
the preceding year.

This reftoration of air, I found, depended
upon the *vegetating ftate* of the plant ; for
though I kept a great number of the frefh
leaves of mint in a fmall quantity of air in
which candles had burned out, and changed
them frequently, for a long fpace of time, I
could perceive no melioration in the ftate of
the air.

E 3   This

This remarkable effect does not depend upon any thing peculiar to *mint*, which was the plant that I always made ufe of till July 1772; for on the 16th of that month, I found a quantity of this kind of air to be perfectly reftored by fprigs of *balm*, which had grown in it from the 7th of the fame month.

That this reftoration of air was not owing to any *aromatic effluvia* of thefe two plants, not only appeared by the *effential oil of mint* having no fenfible effect of this kind; but from the equally complete reftoration of this vitiated air by the plant called *groundfel*, which is ufually ranked among the weeds, and has an offenfive fmell. This was the refult of an experiment made the 16th of July, when the plant had been growing in the burned air from the 8th of the fame month. Befides, the plant which I have found to be the moft effectual of any that I have tried for this purpofe is *fpinach*, which is of quick growth, but will feldom thrive long in water. One jar of burned air was perfectly reftored by this plant in four days, and another in two days. This laft was obferved on the 22d of July.

In general, this effect may be prefumed to have taken place in much lefs time than I have mentioned; becaufe I never chofe to make a trial of the air, till I was pretty fure, from preceding obfervations, that the event which I had expected muft have taken place, if it would fucceed at all ; left, returning back that part of the air on which I made the trial, and which would thereby neceffarily receive a fmall mixture of common air, the experiment might not be judged to be quite fair; though I myfelf might be fufficiently fatisfied with refpect to the allowance that was to be made for that fmall imperfection.

S E C T I O N III.

Of INFLAMMABLE AIR.

I have generally made inflammable air in the manner defcribed by Mr. Cavendifh, in the Philofophical Tranfactions, from iron, zinc, or tin ; but chiefly from the two former metals, on account of the procefs being the leaft troublefome : but when I extracted it from vegetable or animal fubftances, or from coals, I put them into a gun-barrel, to the orifice of which I luted a glafs tube, or the ftem of a tobaccopipe, and to the end of this I tied a flaccid bladder in order to catch the generated air ; or I received the air in a veffel of quickfilver, in the manner reprefented Fig. 7.

E 4   There

There is not, I believe, any vegetable or animal fubftance whatever, nor any mineral fubftance, that is inflammable, but what will yield great plenty of inflammable air, when they are treated in this manner, and urged with a ftrong heat ; but, in order to get the moft air, the heat muft be applied as fuddenly, and as vehemently, as poffible. For, notwithftanding the fame care be taken in luting, and in every other refpect, fix or even ten times more air may be got by a fudden heat than by a flow one, though the heat that is laft applied be as intenfe as that which was applied fuddenly. A bit of dry oak, weighing about twelve grains, will generally yield about a fheep's bladder full of inflammable air with a brifk heat, when it will only give about two or three ounce mea-fures, if the fame heat be applied to it very gradually. To what this difference is owing, I cannot tell. Perhaps the phlogifton being extricated more flowly may not be intirely ex-pelled, but form another kind of union with its bafe; fo that charcoal made with a heat flowly applied fhall contain more phlogifton than that which is made with a fudden heat, It may be worth while to examine the proper-ties of the charcoal with this view.

Inflammable air, when it is made by a quick procefs, has a very ftrong and offenfive fmell,
                                                    from

from whatever fubftance it be generated; but this fmell is of three different kinds, according as the air is extracted from mineral, vegetable, or animal fubftances. The laft is exceedingly fetid; and it makes no difference, whether it be extracted from a bone, or even an old and dry tooth, from foft mufcular flefh; or any other part of the animal. The burning of any fubftance occafions the fame fmell: for the grofs fume which arifes from them, before they flame, is the inflammable air they contain, which is expelled by heat, and then readily ig- nited. The fmell of inflammable air is the very fame, as far as I am able to perceive, from whatever fubftance of the fame kingdom it be extracted. Thus it makes no difference whe- ther it be got from iron, zinc, or tin, from any kind of wood, or, as was obferved before, from any part of an animal.

If a quantity of inflammable air be contained in a glafs veffel ftanding in water, and have been generated very faft, it will fmell even through the water, and this water will alfo foon become covered with a thin film, affuming all the different colours. If the inflammable air have been generated from iron, this matter will appear to be a red okre, or the earth of iron, as I have found by collecting a confi- derable quantity of it; and if it have been generated

nerated from zinc, it is a whitiſh ſubſtance,
which I ſuppoſe to be the calx of the metal.
It likewiſe ſettles to the bottom of the veſſel,
and when the water is ſtirred, it has very much
the appearance of wool.   When water is once
impregnated in this manner, it will continue to
yield this ſcum for a conſiderable time after the
air is removed from it.   This I have often obſerved with reſpect to iron.

Inflammable air, made by a violent efferveſcence, I have obſerved to be much more inflammable than that which is made by a weak
efferveſcence, whether the water or the oil of
vitriol prevailed in the mixture.   Alſo the offenſive ſmell was much ſtronger in the former
caſe than in the latter.   The greater degree of
inflammability appeared by the greater number of ſucceſſive exploſions, when a candle was
preſented to the neck of a phial filled with it.*
It is poſſible, however, that this diminution of
inflammability may, in ſome meaſure, ariſe
from the air continuing ſo much longer in the
bladder when it is made very ſlowly : though

* To try this, after every exploſion, which immediately
follows the preſenting of the flame, the mouth of the phial
ſhould be cloſed (I generally do it with a finger of the
hand in which I hold the phial) for otherwiſe the inflammable air will continue burning, though inviſibly in the
day time, till the whole be conſumed.

I think

I think the difference is too great for this caufe to have produced the whole of it. It may, perhaps, deferve to be tried by a different pro- cefs, without a bladder.

Inflammable air is not thought to be mifci- ble with water, and when kept many months, feems, in general, to be as inflammable as ever. Indeed, when it is extracted from vegetable or animal fubftances, a part of it will be imbibed by the water in which it ftands; but it may be prefumed, that in this cafe, there was a mixture of fixed air extracted from the fubftance along with it. I have indifputable evidence, how- ever, that inflammable air, ftanding long in water, has actually loft all its inflammability, and even come to extinguifh flame much more than that air in which candles have burned out. After this change it appears to be greatly diminifhed in quantity, and it ftill continues to kill animals the moment they are put into it.

This very remarkable fact firft occurred to my obfervation on the twenty-fifth of May 1771, when I was examining a quantity of in- flammable air, which had been made from zinc, near three years before. Upon this, I imme- diately fet by a common quart-bottle filled with inflammable air from iron, and another equal quantity from zinc : and examining them in
the

the beginning of December following, that
from the iron was reduced near one half in
quantity, if I be not greatly miftaken; for I
found the bottle half full of water, and I am
pretty clear that it was full of air when it was
fet by. That which had been produced from
zinc was not altered, and filled the bottle as at
firft.

Another inftance of this kind occurred to
my obfervation on the 19th of June 1772, when
a quantity of air, half of which had been inflammable air from zinc, and half air in which
mice had died, and which had been put together the 30th of July 1771, appeared not
to be in the leaft inflammable, but extinguifhed
flame, as much as any kind of air that I had
ever tried. I think that, in all, I have had
four inftances of inflammable air lofing its inflammability, while it ftood in water.

Though air tainted with putrefaction extinguifhes flame, I have not found that animals
or vegetables putrefying inflammable air render it lefs inflammable. But one quantity of
inflammable air, which I had fet by in May
1771, along with the others above mentioned,
had had fome putrid flefh in it; and this air
had loft its inflammability, when it was examined at the fame time with the other in the December

cember following. The bottle in which this air had been kept, fmelled exactly like very ftrong Harrogate water. I do not think that any perfon could have diftinguifhed them.

I have made plants grow for feveral months in inflammable air made from zinc, and alfo from oak ; but, though the plants grew pretty well, the air ftill continued inflammable. The former, indeed, was not fo highly inflammable as when it was frem made, but the latter was quite as much fo ; and the diminution of inflammability in the former cafe, I attribute to fome other caufe than the growth of the plant,

No kind of air, on which I have yet made the experiment, will conduct electricity; but the colour of an electric fpark is remarkably different in fome different kinds of air, which feems to fhew that they are not equally good non-conductors. In fixed air, the electric fpark is exceedingly white ; but in inflammable air it is of a purple, or red colour. Now, fince the moft vigorous fparks are always the whiteft, and, in other cafes, when the fpark is red, there is reafon to think that the electric matter paffes with difficulty, and with lefs rapidity : it is poffible that the inflammable air may contain particles which conduct electricity, though very imperfectly ; and that the whitenefs of the fpark in the fixed air, may be

owing

owing to its meeting with no conducting particles at all.   When an explofion was made in a
quantity of inflammable air, it was a little white
in the center, but the edges of it were ftill
tinged with a beautiful purple.   The degree of
whitenefs in this cafe was probably owing to the
electric matter rushing with more violence in an
explofion than in a common fpark.

Inflammable air kills animals as fuddenly
as fixed air, and, as far as can be perceived,
in the fame manner, throwing them into convulfions, and thereby occafioning prefent death.
I had imagined that, by animals dying in a quantity of inflammable air, it would in time become lefs noxious ; but this did not appear to
be the cafe ; for I killed a great number of mice
in a fmall quantity of this air ; which I kept feveral months for this purpofe, without its being
at all fenfibly mended ; the laft, as well as the
firft moufe, dying the moment it was put
into it.

I once imagined that, fince fixed and inflammable air are the reverfe of one another, in feveral remarkable properties, a mixture of them
would make common air ; and while I made the
mixtures in bladders, I imagined that I had fucceeded in my attempt ; but I have fince found
that thin bladders do not fufficiently prevent
                                                        the

the air that is contained in them from mixing
with the external air. Alfo corks will not fufficiently confine different kinds of air, unlefs
the phials in which they are confined be fet with
their mouths downwards, and a little water lie
in the necks of them, which, indeed, is equivalent to the air ftanding in veffels immerfed in
water. In this manner, however, I have kept
different kinds of air for feveral years.

Whatever methods I took to promote the
mixture of fixed and inflammable air, they were
all ineffectual. I think it my duty, however,
to recite the iffue of an experiment or two of
this kind, in which equal mixtures of thefe two
kinds of air had ftocd near three years, as they
feem to fhew that they had in part affected one
another, in that long fpace of time. Thefe
mixtures I examined April 27, 1771. One of
them had ftood in quickfilver, and the other in
a corked phial, with a little water in it. On
opening the latter in water, the water inftantly
rufhed in, and filled almoft half of the phial,
and very little more was abforbed afterwards.
In this cafe the water in the phial had probably
abforbed a confiderable part of the fixed air,
fo that the inflammable air was exceedingly rarefied; and yet the whole quantity that muft
have been rendered non-elaftic was ten times
more than the bulk of the water, and it has not
been

been found that water can contain much more
than its own bulk of fixed air.   But in other
caſes I have found the diminution of a quantity
of air, and eſpecially of fixed air, to be much
greater than I could well account for by any
kind of abſorption.

The phial which had ſtood immerſed in quickſilver had loſt very little of its original quantity
of air ; and being now opened in water, and
left there, along with another phial, which was
juſt then filled, as this had been three years be-
rore, viz. with air half inflammable and half
fixed, I obſerved that the quantity of both was
diminiſhed, by the abſorption of the water, in
the ſame proportion.

Upon applying a candle to the mouths of
the phials which had been kept three years,
that which had ſtood in quickſilver went off at
one exploſion, exactly as it would have done
if there had been a mixture of common air
with the inflammable.   As a good deal depends
upon the apertures of the veſſels in which the
inflammable air is mixed, I mixed the two
kinds of air in equal proportions in the ſame
phial, and after letting the phial ſtand ſome
days in water, that the fixed air might be abſorbed, I applied a candle to it, but it made
ten or twelve exploſions (ſtopping the phial
                                                    after

after each of them) before the inflammable matter was exhaufted.

The air which had been confined in the corked phial exploded in the very fame manner as an equal and frefh mixture of the two kinds of air in the fame phial, the experiment being made as foon as the fixed air was abforbed, as before ; fo that in this cafe, the two kinds of air did not feem to have affected one another at all.

Confidering inflammable air as air united to, or loaded with phlogifton, I expofed to it feveral fubftances, which are faid to have a near affinity with phlogifton, as oil of vitriol, and fpirit of nitre (the former for above a month), but without making any fenfible alteration in it.

I obferved, however, that inflammable air, mixed with the fumes of fmoking fpirit of nitre, goes off at one explofion, exactly like a mixture of half common and half inflammable air. This I tried feveral times, by throwing the inflammable air into a phial full of fpirit of nitre, with its mouth immerfed in a bafon containing fome of the fame fpirit, and then applying the flame of a candle to the mouth of the phial, the moment that it was uncovered, after it had been taken out of the bafon.

F This

This remarkable effect I haftily concluded to have arifen from the inflammable air having been in part deprived of its inflammability, by means of the ftronger affinity, which the fpirit of nitre had with phlogifton, and therefore I imagined that by letting them ftand longer in contact, and efpecially by agitating them ftrongly together, I fhould deprive the air of all its inflammability; but neither of thefe operations fucceeded,. for ftill the air was only exploded at once, as before.

And laftly, when I paffed a quantity of inflammable air, which had been mixed with the fumes of fpirit of nitre, through a body of water, and received it in another veffel, it appeared not to have undergone any change at all, for it went off in feveral fucceffive explofions, like the pureft inflammable air. The effect abovementioned muft, therefore, have been owing to the fumes of the fpirit of nitre fupplying the place of common air for the purpofe of ignition, which is analogous to other experiments with nitre.

Having had the curiofity, on the 25th of July 1772, to expofe a great variety of different kinds of air to water out of which the air it contained had been boiled, without any particular view; the refult was, in feveral refpects, altogether

gether unexpected, and led to a variety of new
obfervations on the properties and affinities of
feveral kinds of air with refpect to water.
Among the reft three fourths of that which
was inflammable was abforbed by the water in
about two days, and the remainder was inflammable, but weakly fo.

Upon this, I began to agitate a quantity of
ftrong inflammable air in a glafs jar, ftanding
in a pretty large trough of water, the furface
of which was expofed to the common air, and
I found that when I had continued the operation about ten minutes, near one fourth of the
quantity of air had difappeared; and finding
that the remainder made an effervefcence with
nitrous air, I concluded that it muft have become fit for refpiration, whereas this kind of
air is, at the firft, as noxious as any other kind
whatever. To afcertain this, I put a moufe
into a veffel containing $2\frac{1}{4}$ ounce meafures of
it, and obferved that it lived in it twenty minutes, which is as long as a moufe will generally live in the fame quantity of common air.
This moufe was even taken out alive, and recovered very well. Still alfo the air in which
it had breathed fo long was inflammable, though
very weakly fo. I have even found it to be fo
when a moufe has actually died in it. Inflammable air thus diminifhed by agitation in

water, makes but one explofion on the approach
of a candle, exactly like a mixture of inflammable air with common air.

From this experiment I concluded that, by
continuing the fame procefs, I fhould deprive
inflammable air of all its inflammability, and
this I found to be the cafe; for, after a longer
agitation, it admitted a candle to burn in it,
like common air, only more faintly; and indeed by the teft of nitrous air it did not appear
to be near fo good as common air. Continuing
the fame procefs ftill farther, the air which had
been moft ftrongly inflammable a little before,
came to extinguifh a candle, exactly like air in
which a candle had burned out, nor could they
be diftinguifhed by the teft of n trous air.

I found, by repeated trials, that it was difficult to catch the time in which inflammable air
obtained from metals, in coming to extinguifh
flame, was in the ftate of common air, fo that
the tranfition from rhe one to the other muft be
very fhort.   Indeed I think that in many, perhaps in moft cafes, there may be no proper medium at all, the phlogifton paffing at once from
that mode of union with its bafe which conftitutes inflammable air, to that which conftitutes
an air that extinguifhes flame, being fo much
overloaded as to admit of no more.   I readily,
                                    how-

however, found this middle ftate in a quantity of inflammable air extracted from oak, which air I had kept a year, and in which a plant had grown, though very poorly, for fome part of the time. A quantity of this air, after being agitated in water till it was diminifhed about one half, admitted a candle to burn in it exceedingly well, and was even hardly to be diftinguifhed from common air by the teft of nitrous air.

I took fome pains to afcertain the quantity of diminution, in frefh made and very highly-inflammable air from iron, at which it ceafed to be inflammable, and, upon the whole, I concluded that it was fo when it was diminifhed a little more than one half; for a quantity which was diminifhed exactly one half had fomething inflammable in it, but in the flighteft degree imaginable. It is not improbable, however, but there may be great differences in the refult of this experiment.

Finding that water would imbibe inflammable air, I endeavoured to impregnate water with it, by the fame procefs by which I had made water imbibe fixed air ; but though I found that diftilled water would imbibe about one fourteenth of its bulk of inflammable air, I could not perceive that the tafte of it was fenfibly altered.

F 3   S E C T.

SECTION IV.

Of Air *infeꝗted with* ANIMAL RESPIRATION, *or* PUTREFACTION.

That candles will burn only a certain time, in a given quantity of air is a fact not better known, than it is that animals can live only a certain time in it ; but the cauſe of the death of the animal is not better known than that of the extinction of flame in the ſame circumſtances ; and when once any quantity of air has been rendered noxious by animals breathing in it as long as they could, I do not know that any methods have been diſcoyered of rendering it 'fit for breathing again. It is evident, however, that there muſt be ſome proviſion in nature for this purpoſe, as well as for that of rendering the air fit for ſuſtaining flame ; for without it the whole maſs of the atmoſphere would, in time, become unfit for the purpoſe of animal life ; and yet there is no reaſon to think that it is, at preſent, at all leſs fit for reſpiration than it has ever been. I flatter myſelf, however, that I have hit upon two of the methods employed by nature for this great purpoſe. How many others there may be, I cannot teil.

When

When animals die upon being put into air in which other animals have died, after breathing in it as long as they could, it is plain that the cause of their death is not the want of any *pabulum vitæ*, which has been fuppofed to be contained in the air, but on account of the air being impregnated with fomething ftimulating to their lungs; for they almoft always die in convulfions, and are fometimes affected fo fuddenly, that they are irrecoverable aftei a fingle infpiration, though they be withdrawn immediately, and every method has been taken to bring them to life again. They are affected in the fame manner, when they are killed in any other kind of noxious air that I have tried, viz. fixed air, inflammable air, air filled with the fumes of brimftone, infected with putrid matter, in which a mixtuie of iron filings and brimftone has ftood, or in which charcoal has been burned, or metals calcined, or in nitrous air, &c.

As it is known that *convulfions* weaken, and exhauft the vital powers, much more than the moft vigorous *voluntary* action of the mufcles, perhaps thefe univerfal convulfions may exhauft the whole of what we may call the *vis vitæ* at once, at leaft that the lungs may be rendered abfolutely incapable of action, till the animal be

fuffo

ſuffocated, or be irrecoverable for want of reſpiration.

If a mouſe (which is an animal that I have
commonly made uſe of for the purpoſe of theſe
experiments) can ſtand the firſt ſhock of this
ſtimulus, or has been habituated to it by degrees, it will live a conſiderable time in air in
which other mice will die inſtantaneouſly. I
have frequently found that when a number of
mice have been confined in a given quantity of
air, leſs than half the time that they have actually lived in it, a freſh mouſe being introduced to them has been inſtantly thrown into
convulſions, and died. It is evident, therefore,
that if the experiment of the Black Hole were
to be repeated, a man would ſtand the better
chance of ſurviving it, who ſhould enter at the
firſt, than at the laſt hour.

I have alſo obſerved, that young mice will
always live much longer than old ones, or than
thoſe which are full grown, when they are confined in the ſame quantity of air. I have ſometimes known a young mouſe to live ſix hours in
the ſame circumſtances in which an old mouſe
has not lived one. On theſe accounts, experiments with mice, and, for the ſame reaſon, no
doubt, with other animals alſo, have a conſiderable degree of uncertainty attending them ; and
                                    therefore,

therefore, it is neceffary to repeat them frequently, before the refult can be abfolutely depended upon. But every perfon of feeling will
rejoice with me in the difcovery of *nitrous air,*
to be mentioned hereafter, which fuperfedes
many experiments with the refpiration of animals, being a much more accurate teft of the
purity of air.

The difcovery of the provifion in nature for
reftoring air, which has been injured by the refpiration of animals, having long appeared to
me to be one of the moft important problems in
natural philofophy, I have tried a great variety
of fchemes in order to effeCt it. In thefe my guide
has generally been to confider the influences to
which the atmofphere is, in faCt, expofed; and,
as fome of my unfuccefsful trials may be of ufe
to thofe who are difpofed to take pains in the
farther inveftigation of this fubjeCt, I fhall mention the principal of them.

The noxious effluvium with which air is
loaded by animal refpiration, is not abforbed by
ftanding, without agitation; in frefh or falt water. I have kept it many months in frefh water,
when, inftead of being meliorated, it has feemed
to become even more deadly, fo as to require
more time to reftore it, by the methods which
will be explained hereafter, than air which has
been

been lately made noxious. I have even ſpent
ſeveral hours in pouring this air from one glaſs
veſſel into another, in water, ſometimes as cold,
and ſometimes as warm, as my hands could
bear, it, and have ſometimes alſo wiped the
veſſels many times, during the courſe of the experiment, in order to take off that part of the
noxious matter, which might adhere to the glaſs
veſſels, and which evidently gave them an offenſive ſmell ; but all theſe methods were generally without any ſenſible effect. The *motion*,
alſo, which the air received in theſe circumſtances, it is very evident, was of no uſe for
this purpoſe. I had not then thought of the
ſimple, but moſt effectual method of agitating
air in water, by putting it into a tall jar and
ſhaking it with my hand.

This kind of air is not reſtored by being expoſed to the *light*, or by any other influence to
which it is expoſed, when confined in a thin
phial, in the open air, for ſome months.

Among other experiments, I tried a great variety of different *effluvia*, which are continually
exhaling into the air, eſpecially of thoſe ſubſtances which are known to reſiſt putrefaction ;
but I could not by theſe means effect any melioration of the noxious quality of this kind of
air.

Having

Having read, in the memoirs of the Imperial Society, of a plague not affecting a particular village, in which there was a large fulphur-work, I immediately fumigated a quantity of this kind of air ; or (which will hereafter appear to be the very fame thing) air tainted with putrefaction, with the fumes of burning brim-ftone, but without any effect.

I once imagined, that the *nitrous acid* in the air might be the general reftorative which I was in queft of ; and the conjecture was favoured, by finding that candles would burn in air extracted from faltpetre. I therefore fpent a good deal of time in attempting, by a burning glafs, and other means, to impregnate this noxious air, with fome effluvium of faltpetre, and, with the fame view, introduced into it the fumes of the fmoaking fpirit of nitre ; but both thefe methods were altogether ineffectual.

In order to try the effect of *heat*, I put a quantity of air, in which mice had died, into a bladder, tied to the end of the ftem of a tobacco-pipe, at the other end of which was another bladder, out of which the air was carefully preffed. I then put the middle part of the ftem into a chafing-difh of hot coals, ftrongly urged with a pair of bellows ; and, preffing the bladders alternately, I made the air pafs feveral
times

times through the heated part of the pipe.   I
have alfo made this kind of air very hot, ftanding in water before the fire.   But neither of
thefe methods were of any ufe.

Rarefaction and *condenfation* by inftruments
were alfo tried, but in vain.

Thinking it poffible that the *earth* might imbibe the noxious quality of the air, and thence
fupply the roots of plants with fuch putrefcent
matter as is known to be nutritive to them, I
kept a quantity of air, in which mice had died,
in a phial, one half of which was filled with fine
garden-mould ; but, though it ftood two
months in thefe circumftances, it was not the
better for it.

I once imagined that, fince feveral kinds of
air cannot be long feparated from common air,
by being confined in bladders, in bottles well
corked, or even clofed with ground ftopples, the
affinity between this noxious air and the common
air might be fo great, that they would mix
through a body of water interpofed between
them ; the water continually receiving from the
one, and giving to the other, efpecially as water receives fome kind of impregnation from,
I believe, every kind of air to which it is contiguous ;

guous; but I have feen no reafon to conclude, that a mixture of any kind of air with the common air can be produced in this manner.

I have kept air in which mice have died, air in which candles have burned out, and inflammable air, feparated from the common air, by the flighteft partition of water that I could well make, fo that it might not evaporate in a day or two, if I fhould happen not to attend to them ; but I found no change in them after a month or fix weeks. The inflammable air was ftill inflammable, mice died inftantly in the air in which other mice had died before, and candles would not burn where they had burned out before.

Since air tainted with animal or vegetable putrefaction is the fame thing with air rendered noxious by animal refpiration, I fhall now recite the obfervations which I have made upon this kind of air, before I treat of the method of reftoring them.

That thefe two kinds of air are, in fact, the fame thing, I conclude from their having feveral remarkable common properties, and from their differing in nothing that I have been able to obferve. They equally extinguifh flame, they are equally noxious to animals, they are equally,

equally, and in the fame way, offenfive to the
fmell, and they are reftored by the fame means.

Since air which has paffed through the lungs
is the fame thing with air tainted with animal
putrefaction, it is probable that one ufe of the
lungs is to carry off a *putrid effluvium*, without
which, perhaps, a living body might putrefy
as foon as a dead one.

When a moufe putrefies in any given quantity of air, the bulk of it is generally increafed
for a few days; but in a few days more it begins to fhrink up, and in about eight or ten
days, if the weather be pretty warm, it will be
found to be diminifhed $\frac{1}{6}$, or $\frac{1}{7}$ of its bulk,
If it do not appear to be diminifhed after this
time, it only requires to be paffed through water, and the diminution will not fail to be fenfible. I have fometimes known almoft the
whole diminution to take place, upon once or
twice paffing through the water. The fame is
the cafe with air, in which animals have
breathed as long as they could. Alfo, air in
which candles have burned out may almoft always be farther reduced by this means.

All thefe proceffes, as I obferved before, feem
to difpofe the compound mafs of air to part with
                                                fome

fome conftituent part belonging to it (which appears to be the *fixed air* that enters into its conftitution) and this being mifcible with water,
muft be brought into contaɛt with it, in order
to mix with it to the moft advantage, efpecially
when its union with the other conftituent principles of the air is but partially broken.

I have put mice into veffels which had their
mouths immerfed in quickfilver, and obferved
that the air was not much contraɛted after they
were dead or cold; but upon withdrawing the
mice, and admitting lime water to the air, it
immediately became turbid, and was contraɛted
in its dimenfions as ufual.

I tried the fame thing with air tainted with
putrefaɛtion, putting a dead moufe to a quantity of common air, in a veffel which had its
mouth immerfed in quickfilver, and after a week
I took the moufe out, drawing it through the
quickfilver, and obferved that, for fome time,
there was an apparent increafe of the air perhaps about $\frac{1}{20}$. After this, it ftood two days
in the quickfilver, without any fenfible alteration; and then admitting water to it, it began
to be abforbed, and continued fo, till the original quantity was diminifhed about $\frac{1}{5}$. If, inftead of common water, I had made ufe of lime-
water

water in this experiment, I make no doubt but
it would have become turbid.

If a quantity of lime-water in a phial be put
under a glafs veffel ftanding in water, it will
not become turbid, and provided the accefs of
the common air be prevented, it will continue
lime-water, I do not know how long; but if
a moufe be left to putrefy in the veffel, the
water will depofit all its lime in a few days.
This is owing to the fixed air depofited by the
common air, and perhaps alfo from more fixed
air difcharged from the putrefying fubftances
in fome part of the procefs of putrefaction.

The air that is difcharged from putrefying
fubftances feems, in fome cafes, to be chiefly
fixed air, with the addition of fome other effluvium, which has the power of diminifhing common air. The refemblance between the true
putrid effluvium and fixed air in the following
experiment, which is as decifive as I can poffibly contrive it, appeared to be very great;
indeed much greater than I had expected. I
put a dead moufe into a tall glafs veffel, and
having filled the remainder with quickfilver,
and fet it, inverted, in a pot of quickfilver, I
let it ftand about two months, in which time
the putrid effluvium iffuing from the moufe
had filled the whole veffel, and part of the dif-
foived

folved blood, which lodged upon the furface of the quickfilver, began to be thrown
out. I then filled another glafs veffel, of
the fame fize and fhape, with as pure fixed
air as I could make, and expofed them both,
at the fame time, to a quantity of lime-water.
In both cafes the water grew turbid alike,
it rofe equally faft in both the veffels, and likewife equally high ; fo that about the fame
quantity remained unabforbed by the water.
One of thefe kinds of air, however, was exceedingly fweet and pleafant, and the other in-
fufferably offenfive ; one of them alfo would
have made an addition to any quantity of common air, with which it had been mixed, and the
other would have diminifhed it. This, at leaft,
would have been the confequence, if the moufe
itfelf had putrefied in any quantity of common air.

It feems to depend, in fome meafure, upon
the *time,* and other circumftances, in the diffo-
lution of animal or vegetable fubftances, whether they yield the proper putrid effluvium, or
fixed, or inflammable air ; but the experiments
which I have made upon this fubject, have not
been numerous enough to enable me to decide
with certainty concerning thofe circumftances.

Putrid cabbage, green or boiled, infects
the air in the very fame manner as putrid

G   animal

animal fubftances. Air thus tainted is equally
contracted in its dimenfions, it equally extinguifhes flame, and is equally noxious to animals; but they affect the air very differently,
if the heat that is applied to them be confiderable.

If beef or mutton, raw or boiled, be placed
fo near to the fire, that the heat to which it is
expofed fhall equal, or rather exceed, that of
the blood, a confiderable quantity of air will be
generated in a day or two, about ½th of which I
have generally found to be abforbed by water,
while all the reft was inflammable; but air
generated from vegetables, in the fame circumftances, will be almoft all fixed air, and no part
of it inflammable. This I have repeated again
and again, the whole procefs being in quickfilver; fo that neither common air nor water,
had any accefs to the fubftance on which the
experiment was made ; and the generation of
air, or effluvium of any kind, except what
might be abforbed by quickfilver, or reforbed
by the fubftance itfelf, might be diftinctly
noted.

A vegetable fubftance, after ftanding a day
or two in thefe circumftances, will yield nearly
all the air that can be extracted from it, in that
degree of heat; whereas an animal fubftance
will

will continue to give more air, or effluvium, of
fome kind or other, with very little alteration,
for many weeks. It is remarkable, however,
that though a piece of beef òr muttoñ, plunged
in quickfilver, and kept in this degree of heat,
yield air, the bulk of which is inflammable,
and contraĉts no putrid fmell (at leaft, in a day
or two) a moufe treatéd in the fame manner,
yields the proper putrid effluvium, as indeed
the fmell fufficiently indicatesِ

That the putrid effluvium will mix with
water feems to be evident from the following
experiment. If a moufe be put into a jar full
of water, ftanding with its mouth inverted in
another veffel of water, a confiderable quantity
of elaftic matter (and which may, therefore,
be called *air*) will foon be generated, unlefs the
weather be fo cold as to check all putrefaĉtion.
After a fhort time, the water contraĉts an extremely fetid and offenfive fmell, which feems
to indicate that the putrid effluvium pervadeś
the water, and affeĉts the neighbouring air;
and fince, after this, there is often no increafe
of the air, that feems to be the very fubftance
which is carried off through the water, as faft
as it is generated; and the offenfive fmell is a
fufficient proof that it is not fixed air. For
this has a very agreeable flavour, whether it be
produced by fermentation, or extraĉted from

G 2   chalk

chalk by oil of vitriol ; affecting not only the mouth, but even the noftrils, with a pungency which is peculiarly pleafing to a certain degree, as any perfon may eafily fatisfy himfelf, who will chufe to make the experiments.

If the water in which the moufe was im-merfed, and which is faturated with the putrid air, be changed, the greater part of the putrid air, will, in a day or two, be abforbed, though the moufe continues to yield the putrid efflu-vium as before ; for as foon as this frefh water becomes faturated with it, it begins to be offen-five to the fmell, and the quantity of the putrid air upon its furface increafes as before. I kept a moufe producing putrid air in this manner for the fpace of feveral months.

Six ounce meafures of air not readily abforbed by water, appeared to have been generated from one moufe, which had been putrefying eleven days in confined air, before it was put into a jar which was quite filled with water, for the purpofe of this obfervation.

Air thus generated from putrid mice ftanding in water, without any mixture of common air, extinguifhes flame, and is noxious to animals, but not more fo than common air only tainted with putrefaction. It is exceedingly difficult
and

and tedious to collect a quantity of this putrid air, not mifcible in water, fo very great a proportion of what is collected being abforbed by the water in which it is kept ; but what that proportion is, I have not endeavoured to afcertain. It is probably the fame proportion that that part of fixed air, which is not readily abforbed by water, bears to the reft ; and therefore this air, which I at firft diftinguifhed by the name of *the putrid effluvium*, is probably the fame with fixed air, mixed with the phlogiftic matter, which, in this and other proceffes, diminifhes common air.

Though a quantity of common air be diminifhed by any fubftance putrefying in it, I have not yet found the fame effect to be produced by a mixture of putrid air with common air ; but, in the manner in which I have hitherto made the experiment, I was obliged to let the putrid air pafs through a body of water, which might inftantly abforb the phlogiftic matter that diminifhed the common air.

Infects of various kinds live perfectly well in air tainted with animal or vegetable putrefaction, when a fingle infpiration of it would have inftantly killed any other animal. I have frequently tried the experiment with flies and butterflies. The *aphides* alfo will thrive as well

upon

upon plants growing in this kind of air, as in
the open air. I have even been frequently
obliged to take plants out of the putrid air in
which they were growing, on purpofe to brufh
away the fwarms of thefe infects which infected
them ; and yet fo effectually did fome of them
conceal themfelves, and fo faft did they mul-
tiply, in thefe circumftances, that I could feldom keep the plants quite clear of them.

When air has been frefhly and ftrongly
tainted with putrefaction, fo as to fmell through
the water, fprigs of mint have prefently died,
upon being put into it, their leaves turning
black ; but if they do not die prefently, they
thrive in a moft furprizing manner. In no
other circumftances have I ever feen vegetation
fo vigorous as in this kind of air, which is
immediately fatal to animal life. Though thefe
plants have been crouded in jars filled with this
air, every leaf has been full of life ; frefh fhoots
have branched out in various directions, and have
grown much fafter than other fimilar plants,
growing in the fame expofure in common air.

This obfervation led me to conclude, that
plants, inftead of affecting the air in the fame
manner with animal refpiration, reverfe the
effects of breathing, and tend to keep the atmofphere fweet and wholefome, when it is become
noxious

noxious, in confequence of animals either living and breathing, or dying and putrefying in it.

In order to afcertain this, I took a quantity of air, made thoroughly noxious, by mice breathing and dying in it, and divided it into two parts ; one of which I put into a phial immerfed in water; and to the other (which was contained in a glafs jar, ftanding in water) I put a fprig of mint. This was about the beginning of Auguft 1771, and after eight or nine days, I found that a moufe lived perfeҿly well in that part of the air, in which the fprig of mint had grown, but died the moment it was put into the other part of the fame original quantity of air; and which I had kept in the very fame expofure, but without any plant growing in it.

This experiment I have feveral times repeated ; fometimes ufing air in which animals had breathed and died, and at other times ufing air, tainted with vegetable or animal putrefaction ; and generally with the fame fuccefs.

Once, I let a moufe live and die in a quantity of air which had been noxious, but which had been reftored by this procefs, and it lived nearly as long as I conjeҿured it might have done in an equal quantity of frefh air ; but this is fo exceedingly various, that it is not eafy to form

any

any judgment from it; and in this cafe the
fymptom of *difficult refpiration* feemed to begin
earlier than it would have done in common air.

Since the plants that I made ufe of manifeftly
grow and thrive in putrid air; fince putrid
matter is well known to afford proper nourifhment for the roots of plants; and fince it is
likewife certain that they receive nourifhment
by their leaves as well as by their roots, it
feems to be exceedingly probable, that the
putrid effluvium is in fome meafure extracted
from the air, by means of the leaves of plants,
and therefore that they render the remainder
more fit for refpiration.

Towards the end of the year fome experiments of this kind did not anfwer fo well as they
had done before, and I had inftances of the re-
lapfing of this reftored air to its former noxious
ftate. I therefore fufpended my judgment concerning the efficacy of plants to reftore this kind
of noxious air, till I fhould have an opportunity
of repeating my experiments, and giving more
attention to them. Accordingly I refumed the
experiments in the fummer of the year 1772,
when I prefently had the moft indifputable
proof of the reftoration of putrid air by vegetation; and as the fact is of fome importance,
and the fubfequent variation in the ftate of this
                                                    kind

kind of air is a little remarkable, I think it neceſſary to relate ſome of the facts pretty cir-
cumſtantially.

The air, on which I made the firſt experiments, was rendered exceedingly noxious by
mice dying in it on the 20th of June. Into a
jar nearly filled with one part of this air, I put
a ſprig of mint, while I kept another part of
it in a phial, in the ſame expoſure ; and on the
27th of the ſame month, and not before, I
made a trial of them, by introducing a mouſe
into a glaſs veſſel, containing 2 ½ ounce meaſures filled with each kind of air ; and I noted
the following facts.

When the veſſel was filled with the air in
which the mint had grown, a very large mouſe
lived five minutes in it, before it began to ſhew
any ſign of uneaſineſs. I then took it out,
and found it to be as ſtrong and vigorous as
when it was firſt put in ; whereas in that air
which had been kept in the phial only, without
a plant growing in it, a younger mouſe continued not longer than two or three ſeconds,
and was taken out quite dead. It never breathed
after, and was immediately motionleſs. After
half an hour, in which time the larger mouſe
(which I had kept alive, that the experiment
might be made on both the kinds of air with
the very ſame animal) would have been ſufficiently

ciently recruited, fuppofing it to have received
any injury by the former experiment, was put
into the fame veffel of air; but though it was
withdrawn again, after being in it hardly one
fecond, it was recovered with difficulty, not
being able to ftir from the place for near a
minute. After two days, I put the fame moufe
into an equal quantity of common air, and
obferved that it continued feven minutes without any fign of uneafinefs; and being very
uneafy after three minutes longer, I took it out.
Upon the whole, I concluded that the reftored
air wanted about one fourth of being as wholefome as common air.   The fame thing alfo appeared when I applied the teft of nitrous air.

In the feven days, in which the mint was
growing in this jar of noxious air, three old
fhoots had extended themfelves about three
inches, and feveral new ones had made their
appearance in the fame time. Dr. Franklin and
Sir John Pringle happened to be with me, when
the plant had been three or four days in this
ftate, and took notice of its vigorous vegetation, and remarkably healthy appearance in
that confinement.

On the 30th of the fame month, a moufe
lived fourteen minutes, breathing naturally all
the time, and without appearing to be much
uneafy,

uneasy, till the last two minutes, in the veffel containing two ounce meafures and a half of air which had been rendered noxious by mice breathing in it almoft a year before, and which I had found to be moft highly noxious on the 19th of this month, a plant having grown in it, but not exceedingly well, thefe eleven days; on which account I had deferred making the trial fo long. The reftored air was affected by a mixture of nitrous air, almoft as much as common air.

As this putrid air was thus eafily reftored to a confiderable degree of fitnefs for refpiration, by plants growing in it, I was in hopes that by the fame means it might in time be fo much more perfectly reftored, that a candle would burn it it; and for this purpofe I kept plants growing in the jars which contained this air till the middle of Auguft following, but did not take fufficient care to pull out all the old and rotten leaves. The plants, however, had grown, and looked fo well upon the whole, that I had no doubt but that the air muft conftantly have been in a mending ftate; when I was exceedingly furprized to find, on the 24th of that month, that though the air in one of the jars had not grown worfe, it was no better; and that the air in the other jar was fo much worfe than it had been, that a moufe would have died in it in a

few

few feconds. It alfo made no effervefcence with nitrous air, as it had done before.

Sufpecting that the fame plant might bo capable of reftoring putrid air to a certain degree only, or that plants might have a contrary tendency in fome ftages of their growth, I withdrew the old plant, and put a frefh one in its place; and found that, after feven days, the air was reftored to its former wholefome ftate. This fact I confider as a very remarkable one, and well deferving of a farther inveftigation, as it may throw more light upon the principles of vegetation. It is not, however, a fingle fact; for I had feveral inftances of the fame kind in the preceding year; but it feemed fo very extraordinary, that air fhould grow worfe by the continuance of the fame treatment by which it had grown better, that, whenever I obferved it, I concluded that I had not taken fufficient care to fatisfy myfelf of its previous reftoration.

That plants are capable of perfectly reftoring air injured by refpiration, may, I think, be inferred with certainty from the perfect reftoration, by this means, of air which had paffed through my lungs, fo that a candle would burn in it again, though it had extinguifhed flame before, and a part of the fame original quantity
of

of air ftill continued to do fo. Of this one
inftance occurred in the year 1771, a fprig of
mint having grown in a jar of this kind of air,
from the 25th of July to the 17th of Auguft
following; and another trial I made, with the
fame fuccefs, the 7th of July 1772, the plant
having grown in it from the 29th of June preceding. In this cafe alfo I found that the effect
was not owing to any virtue in the leaves of
mint; for I kept them conftantly changed in
a quantity of this kind of air, for a confiderable time, without making any fenfible alteration in it.

These proofs of a partial reftoration of air
by plants in a ftate of vegetation, though in a
confined and unnatural fituation, cannot but
render it highly probable, that the injury which
is continually done to the atmofphere by the
refpiration of fuch a number of animals, and
the putrefaction of fuch maffes of both vegetable
and animal matter, is, in part at leaft, repaired
by the vegetable creation. And, notwithftanding the prodigious mafs of air that is corrupted
daily by the above-mentioned caufes; yet, if
we confider the immenfe profufion of vegetables
upon the face of the earth, growing in places
fuited to their nature, and confequently at full
liberty to exert all their powers, both inhaling
and exhaling, it can hardly be thought, but
that

that it may be a ſufficient counterbalance to it,
and that the remedy is adequate to the evil.

Dr. Franklin, who, as I have already obſerved, ſaw ſome of my plants in a very flouriſhing
ſtate, in highly noxious air, was pleaſed to ex-
preſs very great ſatisfaction with the reſult of
the experiments. In his anſwer to the letter in
which I informed him of it, he ſays,

" That the vegetable creation ſhould reſtore
" the air which is ſpoiled by the animal part of
" it, looks like a rational ſyſtem, and ſeems to
" be of a piece with the reſt. Thus fire puri-
" fies water all the world over. It purifies it
" by diſtillation, when it raiſes it in vapours,
" and lets it fall in rain ; and farther ſtill by fil-
" tration, when, keeping it fluid, it ſuffers that
" rain to percolate the earth. We knew be-
" fore that putrid animal ſubſtances were con-
" verted into ſweet vegetables, when mixed with
" the earth, and applied as manure ; and now,
" it ſeems, that the ſame putrid ſubſtances,
" mixed with the air, have a ſimilar effect.
" The ſtrong thriving ſtate of your mint in pu-
" trid air ſeems to ſhew that the air is mended
" by taking ſomething from it, and not by ad-
" ding to it." He adds, " I hope this will
" give ſome check to the rage of deſtroying
" trees that grow near houſes, which has
                                        " ac-

" accompanied our late improvements in
" gardening, from an opinion of their being
" unwholefome. I am certain, from long ob-
" fervation, that there is nothing unhealthy in
" the air of woods ; for we Americans have
" every where our country habitations in the
" midft of woods, and no people on earth en-
" joy better health, or are more prolific."

Having rendered inflammable air perfectly innoxious by continued *agitation in a trough of
water,* deprived of its air, I concluded that
other kinds of noxious air might be reftored by
the fame means ; and I prefently found that this
was the cafe with putrid air, even of more than
a year's ftanding. I fhall obferve once for all,
that this procefs has never failed to reftore any
kind of noxious air on which I have tried it,
viz. air injured by refpiration or putrefaction,
air infected with the fumes of burning charcoal, and of calcined metals, air in which a
mixture of iron filings and brimftone, that in
which paint made of white lead and oil has
ftood, or air which has been diminifhed by a
mixture of nitrous air. Of the remarkable effect which this procefs has on nitrous air itfelf,
an account will be given in its proper place.

If this procefs be made in water deprived of
air, either by the air-pump, by boiling, or by diftillation,

tillation, or if frefh rain-water be ufed, the air
will always be diminifhed by the agitation ; and
this is certainly the faireft method of making
the experiment. If the water be frefh pump-
water, there will always be an increafe of the
air by agitation, the air contained in the water
being fet loofe, and joining that which is in the
jar. In this cafe, alfo, the air has never failed
to be reftored ; but then it might be fufpected
that the melioration was produced by the addition of fome more wholefome ingredient. As
thefe agitations were made in jars with wide
mouths, and in a trough which had a large furface expofed to the common air, I take it for
granted that the noxious effluvia, whatever they
be, were firft imbibed by the water, and thereby tranfmitted to the common atmofphere. In
fome cafes this was fufficiently indicated by the
difagreeable fmell which attended the operation.

After I had made thefe experiments, I was
informed that an ingenious phyfician and philofopher had kept a fowl alive twenty-four hours,
in a quantity of air in which another fowl of the
fame fize had not been able to live longer than
an hour, by contriving to make the air, which
it breathed, pafs through no very large quantity of acidulated water, the furface of which
was not expofed to the common air ; nd that
even

even when the water was not acidulated, the fowl lived much longer than it could have done, if the air which it breathed had not been drawn through the water.

As I should not have concluded that this experiment would have succeeded so well, from any observations that I had made upon the subject, I took a quantity of air in which mice had died, and agitated it very strongly, first in about five times its own quantity of distilled water, in the manner in which I had impregnated water with fixed air; but though the operation was continued a long time, it made no sensible change in the properties of the air. I also repeated the operation with pump-water, but with as little effect. In this case, however, though the air was agitated in a phial, which had a narrow neck, the surface of the water in the bason was considerably large, and exposed to the common atmosphere, which must have tended a little to favour the experiment.

In order to judge more precisely of the effect of these different methods of agitating air, I transferred the very noxious air, which I had not been able to amend in the least degree by the former method, into an open jar, standing in a trough of water; and when I had agitated it till it was diminished about one third, I found

H it

it to be better than air in which candles had
burned out, as appeared by the teft of the nitrous air ; and a moufe lived in $2\frac{1}{2}$ ounce meafures of it a quarter of an hour, and was not
fenfibly affected the firft ten or twelve minutes.

In order to determine whether the addition of
any *acid* to the water, would make it more capable of reftoring putrid air, I agitated a quantity of it in a phial containing very ftrong vinegar ; and after that in *aqua fortis*, only half
diluted with water ; but by neither of thefe
proceffes was the air at all mended, though the
agitation was repeated, at intervals, during a
whole day, and it was moreover allowed to
ftand in that fituation all night.

Since, however, water in thefe experiments
muft have imbibed and retained a certain portion of the noxious effluvia, before they could
be tranfmitted to the external air, I do not think
it improbable but that the agitation of the fea
and large lakes may be of fome ufe for the purification of the atmofphere, and the putrid
matter contained in water may be imbibed by
aquatic plants, or be depofited in fome other
manner.

Having found, by feveral experiments above-
mentioned, that the proper putrid effluvium
                                                    it

is fomething quite diftinct from fixed air, and
finding, by the experiments of Dr. Macbride,
that fixed air corrects putrefaction ; it occured
to me, that fixed air, and air tainted with putrefaction, though equally noxious when feparate, might make a wholefome mixture, the
one correcting the other ; and I was confirmed
in this opinion by, I believe, not lefs than fifty
or fixty inftances, in which air, that had been
made in the higheft degree noxious, by refpiration or putrefaction, was fo far fweetened, by
a mixture of about four times as much fixed
air, that afterwards mice lived in it exceedingly
well, and in fome cafes almoft as long as in
common air. I found it, indeed, to be more
difficult to reftore *old* putrid air by this means ;
but I hardly ever failed to do it, when the two
kinds of air had ftood a long time together ; by
which I mean about a fortnight or three weeks.

The reafon why I do not abfolutely conclude
that the reftoration of air in thefe cafes was the
effect of fixed air, is that, when I made a trial
of the mixture, I fometimes agitated the two
kinds of air pretty ftrongly together, in a
trough of water, or at leaft paffed it feveral
times through water, from one jar to another, that the fuperfluous fixed air might be
abforbed, not fufpecting at that time that the
agitation could have any other effect. But

H 2   having

having ſince found that very violent, and eſpecially long-continued agitation in water, without
any mixture of fixed air, never failed to render
any kind of noxious air in ſome meaſure fit for
reſpiration (and in one particular inſtance the
mere transferring of the air from one veſſel to
another through the water, though for a much
longer time than I ever uſed for the mixtures
of air, was of conſiderable uſe for the ſame
purpoſe) I began to entertain ſome doubt of
the efficacy of fixed air in this caſe. In ſome
caſes alſo the mixture of fixed air had by no
means ſo much effect on the putrid air as,
from the generality of my obſervations, I
ſhould have expected.

I was always aware, indeed, that it might be
ſaid, that, the reſiduum of fixed air not being
very noxious, ſuch an addition muſt contribute
to mend the putrid air ; but, in order to obviate this objection, I once mixed the reſiduum
of as much fixed air as I had found, by a variety of trials, to be ſufficient to reſtore a given
quantity of putrid air, with an equal quantity
of that air, without making any ſenſible melioration of it.

Upon the whole, I am inclined to think that
this proceſs could hardly have ſucceeded ſo
well as it did with me, and in ſo great a number

ber of trials, unless fixed air have fome tendency to correct air tainted with refpiration or
putrefaction; and it is perfectly agreeable to
the analogy of Dr. Macbride's difcoveries, and
may naturally be expected from them, that it
fhould have fuch an effect.

By a mixture of fixed air I have made
wholefome the refiduum of air generated by
putrefaction only, from mice plunged in water.
This, one would imagine, *a priori*, to be the
moft noxious of all kinds of air. For if common air only tainted with putrefaction be
fo deadly, much more might one expect that
air to be fo, which was generated from putrefaction only; but it feems to be nothing more
than common air (or at leaft that kind of fixed
air which is not abforbed by water) tainted
with putrefaction, and therefore requires no
other procefs to fweeten it. In this cafe, however, we feem to have an inftance of the generation of genuine common air, though mixed
with fomething that is foreign to it. Perhaps
the refiduum of fixed air may be another inftance of the fame nature, and alfo the refiduum
of inflammable air, and of nitrous air, efpecially nitrous air loaded with phlogifton, after
long agitation in water.

Fixed

Fixed air is equally diffufed through the whole mafs of any quantity of putrid air with which it is mixed : for dividing the mixture in‑ to two equal parts, they were reduced in the fame proportion by paffing through water. But this is alfo the cafe with fome of the kinds of air which will not incorporate, as inflammable air, and air in which brimftone has burned.

If fixed air tend to correct air which has been injured by animal refpiration or putre‑ faction, *lime kilns*, which difcharge great quan‑ tities of fixed air, may be wholefome in the neighbourhood of populous cities, the atmo‑ fphere of which muft abound with putrid ef‑ fluvia.   I fhould think alfo that phyficians might avail themfelves of the application of fixed air in many putrid diforders, efpecially as it may be fo eafily adminiftered by way of *clyfter*, where it would often find its way to much of the putrid matter.   Nothing is to be apprehended from the diftention of the bowels by this kind of air, fince it is fo readily ab‑ forbed by any fluid or moift fubftance.

Since fixed air is not noxious *per fe*, but, like fire, only in excefs, I do not think it at all hazardous to attempt to *breathe* it.   It is how‑ ever eafily conveyed into the *ftomach*, in na‑ tural or artificial Pyrmont water, in brifkly

fer‑

fermenting liquors, or a vegetable diet. It is even poffible, that a confiderable quantity of fixed air might be imbibed by the abforbing veffels of the fkin, if the whole body, except the head, fhould be fufpended over a veffel of ftrongly-fermenting liquor ; and in fome putrid diforders this treatment might be very falutary. If the body was expofed quite naked, there would be very little danger from the cold in this fituation, and the air having freer accefs to the fkin might produce a greater effect. Being no phyfician, I run no rifk by throwing out thefe random, and perhaps whimfical propofals.*

Having communicated my obfervations on fixed air, and efpecially my fcheme of applying it by way of *clyfter* in putrid diforders, to Mr. Hey, an ingenious furgeon in Leeds a cafe prefently occurred, in which he had an opportunity of giving it a trial ; and mentioning it to Dr. Hird and Dr. Crowther, two phyficians who attended the patient, they approved the fcheme, and it was put in execution ; both by applying the fixed air by way of clyfter, and at the fame time making the patient drink plentifully of liquors ftrongly impregnated with

* Some time after thefe papers were firft printed, I was pleafed to find the fame propofal in *Dr. Alexander's Experimental Effays.*

H 4

it.

it. The event was fuch, that I requefted Mr. Hey to draw up a particular account of the cafe, defcribing the whole of the treatment, that the public might be fatisfied that this new application of fixed air is perfectly fafe, and alfo, have an opportunity of judging how far it had the effect which I expected from it; and as the application is new, and not unpromifing, I fhall fubjoin his letter to me on the fubject, by way of *Appendix* to thefe papers.

When I began my inquires into the properties of different kinds of air, I engaged my friend Dr. Percival to attend to the *medicinal ufes* of them, being fenfible that his knowledge of philofophy as well as of medicine would give him a fingular advantage for this purpofe. The refult of his obfervations I fhall alfo infert in the Appendix.

SECT-

SECTION V.

Of AIR *in which a mixture of* BRIMSTONE *and* FILINGS *of* IRON *has ftood.*

Reading in Dr. Hales's account of his experiments, that there was a great diminution of the quantity of air in which *a mixture of powdered brimftone and filings of iron, made into a pafte with water*, had ftood, i repeated the experiment, and found the diminution greater than I had expected. This diminution of air is made as effectually, and as expeditiouſly, in quickſilver as in water; and it may be meaſured with the greateft accuracy, becauſe there is neither any previous expanſion or increaſe of the quantity of air, and becauſe it is ſom time before this procefs begins to have any ſenſible effect. This diminution of air is various; but I have generally found it to be between one fifth and one fourth of the whole.

Air thus diminiſhed is not heavier, but rather lighter than common air; and though lime-water does not become turbid when it is expoſed to this air, it is probably owing to the formation of a ſelenitic ſalt, as was the caſe with the ſimple burning of brimſtone above-
men-

mentioned. That fomething proceeding from
the brimftone ftrongly affects the water which
is confined in the fame place with this mixture,
is manifeft from the very ftrong fmell that it has
of the volatile fpirit of vitriol.

I conclude that the diminution of air by this
procefs is of the fame kind with the diminution
of it in the other cafes, becaufe when this mixture is put into air which has been previoufly diminifhed, either by the burning of candles, by
refpiration, or putrefaction, though it never
fails to diminifh it fomething more, it is, however, no farther than this procefs alone would
have done it. If a frefh mixture be introduced
into a quantity of air which had been reduced
by a former mixture, it has little or no farther
effect.

I once obferved, that when a mixture of this
kind was taken out of a quantity of air in which
a candle had before burned out, and in which it
had ftood for feveral days, it was quite cold
and black, as it always becomes in a confined
place; but it prefently grew very hot, fmoak-
ed copoufly, and fmelled very offenfively;
and when it was cold, it was brown, like the
ruft of iron.

I once

I once put a mixture of this kind to a quantity of inflammable air, made from iron, by which means it was diminifhed $\frac{1}{9}$ or $\frac{1}{10}$ in its bulk ; but, as far as I could judge, it was ftill as inflammable as ever. Another quantity of inflammable air was alfo reduced in the fame proportion, by a moufe putrefying in it; but its inflammability was not feemingly leffened.

Air diminifhed by this mixture of iron filings and brimftone, is exceedingly noxious to animals, and I have not perceived that it grows any better by keeping in water. The fmell of it is very pungent and offenfive.

The quantity of this mixture which I made ufe of in the preceding experiments, was from two to four ounce meafures ; but I did not perceive, but that the diminution of the quantity of air (which was generally about twenty ounce meafures) was as great with the fmalleft, as with the largeft quantity. How fmall a quantity is neceffary to diminifh a given quantity of air to a *maximum*, I have made no experiments to afcertain.

As foon as this mixture of iron filings with brimftone and water, begins to ferment, it alfo turns black, and begins to fwell, and it continues

tinues to do fo, till it occupies twice as much
fpace as it did at firft.   The force with which
it expands is great; but how great it is I have
not endeavoured to determine.

When this mixture is immerfed in water, it
generates no air, though it becomes black, and
fwells.

S E C T I O N VI.

Of NITROUS AIR.

Ever fince I firft read Dr. Hales's moft excellent *Statical Effays*, I was particularly ftruck
with that experiment of his, of which an account is given, VOL. I. p. 224. and VOL. II.
p. 280. in which common air, and air generated from the Walton pyrites, by fpirit of
nitre, made a turbid red mixture, and in which
part of the common air was abforbed; but I
never expected to have the fatisfaction of feeing
this remarkable appearance, fuppofing it to be
peculiar to that particular mineral.   Happening to mention this fubject to the Hon. Mr.
Cavendifh, when I was in London, in the
fpring of the year 1772, he faid that he did
not imagine but that other kinds of pyrites,
or the metals might anfwer as well, and that
                                             pro-

probably the red appearance of the mixture depended upon the fpirit of nitre only. This encouraged me to attend to the fubject; and having no pyrites, I began with the folution of the different metals in fpirit of nitre, and catching the air which was generated in the folution, I prefently found what I wanted, and a good deal more.

Beginning with the folution of brafs, on the 4th of June 1772, I firft found this remarkable fpecies of air, only one effect of which, was cafually obferved by Dr. Hales; and he gave fo little attention to it, and it has been fo much unnoticed fince his time, that, as far as I know, no name has been given to it. I therefore found myfelf, contrary to my firft refolution, under an abfolute neceffity of giving a name to this kind of air myfelf. When I firft began to fpeak and write of it to my friends, I happened to diftinguifh it by the name of *nitrous air*, becaufe I had procured it by means of fpirit of nitre only; and though I cannot fay that I altogether like the term, neither myfelf nor any of my friends, to whom I have applied for the purpofe, have been able to hit upon a better; fo that I am obliged, after all, to content myfelf with it.

I have

I have found that this kind of air is readily
procured from iron, copper, brafs, tin, filver,
quickfilver, bifmuth, and nickel, by the nitrous acid only, and from gold and the regulus
of antimony by *aqua regia.* The circumftances attending the folution of each of thefe
metals are various, but hardly worth mentioning, in treating of the properties of the air
which they yield ; which, from what metal foever it is extracted, has, as far as I have been
able to obferve, the very fame properties.

One of the moft confpicuous properties of
this kind of air is the great diminution of any
quantity of common air with which it is mixed, attended with a turbid red, or deep orange
colour, and a confiderable heat. The *fmell*
of it, alfo, is very ftrong, and remarkable,
but very much refembling that of fmoking fpirit of nitre.

The diminution of a mixture of this and
common air is not an equal diminution of both
the kinds, which is all that Dr. Hales could
obferve, but of about one fifth of the common air, and as much of the nitrous air as is
neceffary to produce that effect ; which, as I
have found by many trials, is about one half
as much as the original quantity of common air.
For if one meafure of nitrous air be put to two

meɑ-

meafures of common air, in a few minutes
(by which time the effervefcence will be over,
and the mixture will have recovered its tranfparency) there will want about one ninth of
the original two meafures ; and if both the
kinds of air be very pure, the diminution will
ftill go on flowly, till in a day or two, the
whole will be reduced to one fifth lefs than the
original quantity of common air. This farther
diminution, by long ftanding, I had not obferved at the time of the firft publication of thefe
papers.

I hardly know any experiment that is more
adapted to amaze and furprize than this is,
which exhibits a quantity of air, which, as it
were, devours a quantity of another kind of
air half as large as itfelf, and yet is fo far from
gaining any addition to its bulk, that it is confiderably diminifhed by it. If, after this full
faturation of common air with nitrous air, more
nitrous air be put to it, it makes an addition
equal to its own bulk, without producing the
leaft rednefs, or any other vifible effect.

If the fmalleft quantity of common air be
put to any larger quantity of nitrous air,
though the two together will not occupy fo
much fpace as they did feparately, yet the
quantity will ftill be larger than that of the
nitrous air only. One ounce meafure of common

mon air being put to near twenty ounce meafures of nitrous air, made an addition to it of
about half an ounce meafure. This being a
much greater proportion than the diminution
of common air, in the former experiment,
proves that part of the diminution in the
former cafe is in the nitrous air. Befides, it
will prefently appear, that nitrous air is fubject
to a moft remarkable diminution; and as common air, in a variety of other cafes, fuffers a
diminution from one fifth to one fourth, I
conclude, that in this cafe alfo it does not
exceed that proportion, and therefore that the
remainder of the diminution refpects the nitrous air.

In order to judge whether the *water* contributed to the diminution of this mixture of nitrous and common air, I made the whole procefs feveral times in quickfilver, ufing one third
of nitrous, and two thirds of common air, as
before. In this cafe the rednefs continued a
very long time, and the diminution was not fo
great as when the mixtures had been made in
water, there remaining one feventh more than
the original quantity of common air.

This mixture ftood all night upon the quickfilver; and the next morning I obferved that it
was no farther diminifhed upon the admiffion
of

of water to it, nor by pouring it feveral times
through the water, and letting it ftand in water
two days.

Another mixture, which had ftood about fix
hours on the quickfilver, was diminifhed a little
more upon the admiffion of water, but was never
lefs than the original quantity of common air.
In another cafe however, in which the mixture
had ftood but a very fhort time in quickfilver, the
farther diminution, which took place upon the
admiffion of water, was much more confiderable ; fo that the diminution, upon the whole,
was very nearly as great as if the procefs had
been intirely in water.

It is evident from thefe experiments, that
the diminution is in part owing to the abforption by the water ; but that when the
mixture is kept a long time, in a fituation in
which there is no water to abforb any part of
it, it acquires a conftitution, by which it is
afterwards incapable of being abforbed by
water, or rather, there is an addition to the quantity of air by nitrous air produced by the folution of the quickfilver.

It will be feen, in the fecond part of this
work, that, in the decompofition of nitrous
air by its mixture with common air, there is

I   nothing

nothing at hand when the proceſs is made in quickſilver, with which the acid that entered into its compoſition can readily unite.

In order to determine whether the fixed part of common air was depoſited in the diminution of it by nitrous air, I incloſed a veſſel full of lime-water in the jar in which the proceſs was made, but it occaſioned no precipitation of the lime ; and when the veſſel was taken out, after it had been in that ſituation a whole day, the lime was eaſily precipitated by breathing into it as uſual.

But though the precipitation of the lime was not ſenſible in this method of making the experiment, it is ſufficiently ſo when the whole proceſs is made in lime-water, as will be ſeen in the ſecond part of this work; ſo that we have here another evidence of the depoſition of fixed air from common air. I have made no alteration, however, in the preceding paragraph, becauſe it may not be unuſeful, as a caution to future experimenters.

It is exceedingly remarkable that this efferveſcence and diminution, occaſioned by the mixture of nitrous air, is peculiar to common air, or *air fit for reſpiration*; and, as far as I can judge, from a great number of obſervations,

tions, is at leaft very nearly, if not exactly, in proportion to its fitnefs for this purpofe; fo that by this means the goodnefs of air may be diftinguifhed much more accurately than it can be done by putting mice, or any other animals, to breathe in it.

This was a moft agreeable difcovery to me, as I hope it may be an ufeful one to the public; efpecially as, from this time, I had no occafion for fo large a ftock of mice as I had been ufed to keep for the purpofe of thefe experiments, ufing them only in thofe which required to be very decifive; and in thefe cafes I have feldom failed to know beforehand in what manner they would be affected.

It is alfo remarkable that, on whatever account air is unfit for refpiration, this fame teft is equally applicable. Thus there is not the leaft effervefcence between nitrous and fixed air, or inflammable air, or any fpecies of diminifhed air. Alfo the degree of diminution being from nothing at all to more than one third of the whole of any quantity of air, we are, by this means, in poffeffion of a prodigioufly large *fcale*, by which we may diftinguifh very fmall degrees of difference in the goodnefs of air.

I have not attended much to this circumftance, having ufed this teft chiefly for greater

dif-

differences ; but, if I did not deceive myſelf,
I have perceived a real difference in the air of
my ſtudy, after a few perſons have been with
me in it, and the air on the outſide of the houſe.
Alſo a phial of air having been ſent me, from
the neighbourhood of York, it appeared not to
be ſo good as the air near Leeds ; that is, it was
not diminiſhed ſo much by an equal mixture of
nitrous air, every other circumſtance being as
nearly the ſame as I could contrive.   It may
perhaps be poſſible, but I have not yet attempted
it, to diſtinguiſh ſome of the different winds,
or the air of different times of the year, &c. &c.
by this teſt.

By means of this teſt I was able to determine
what I was before in doubt about, *viz.* the *kind*
as well as the *degree* of injury done to air by
candles burning in it.   I could not tell with
certainty, by means of mice, whether it was at
all injured with reſpect to reſpiration ; and yet
if nitrous air may be depended upon for fur-
niſhing an accurate teſt, it muſt be rather more
than one third worſe than common air, and
have been diminiſhed by the ſame general cauſe
of the other diminutions of air.   For when,
after many trials, I put one meaſure of thoroughly putrid and highly noxious air, into the
ſame veſſel with two meaſures of good whole-
ſome air, and into another veſſel an equal quantity,

tity, *viz.* three meafures of air in which a candle had burned out; and then put equal quantities of nitrous air to each of them, the latter was diminifhed rather more than the former.

It agrees with this obfervation, that *burned air* is farther diminifhed both by putrefaction, and a mixture of iron filings and brimftone; and I therefore take it for granted by every other caufe of the diminution of air. It is probable, therefore, that burned air is air fo far loaded with phlogifton, as to be able to extinguifh a candle, which it may do long before it is fully faturated.

Inflammable air with a mixture of nitrous air burns with a green flame. This makes a very pleafing experiment when it is properly conducted. As, for fome time, I chiefly made ufe of *copper* for the generation of nitrous air, I firft afcribed this circumftance to that property of this metal, by which it burns with a green flame; but I was prefently fatisfied that it muft arife from the fpirit of nitre, for the effect is the very fame from which ever of the metals the nitrous air is extracted, all of which I tried for this purpofe, even filver and gold.

I 3   A mix-

A mixture of oil of vitriol and fpirit of nitre in equal proportions diffolved iron, and the produce was nitrous air ; but a lefs degree of fpirit of nitre in the mixture produced air that was inflammable, and which burned with a green flame. It alfo tinged common air a little red, and diminifhed it, though not much.

The diminution of common air by a mixture of nitrous air, is not fo extraordinary as the diminution which nitrous air itfelf is fubject to from a mixture of iron filings and brimftone, made into a pafte with water. This mixture, as I have already obferved, diminifhes common air between one fifth and one fourth, but has no fuch effect upon any kind of air that has been diminifhed, and rendered noxious by any other procefs ; but when it is put to a quantity of nitrous air, it diminifhes it fo much, that no more than one fourth of the original quantity will be left,

The effect of this procefs is generally per-ceived in five or fix hours, about which time the vifible effervefence of the mixture begins; and in a very fhort time it advances fo rapidly, that in about an hour almoft the whole effect will have taken place. If it be fuffered to ftand a day or two longer, the air will ftill be diminifhed farther, but only a very little farther,

in

in proportion to the firſt diminution. The glaſs jar, in which the air and this mixture have been confined, has generally been ſo much heated in this procefs, that I have not been able to touch it.

Nitrous air thus diminiſhed has not ſo ſtrong a ſmell as nitrous air itſelf, but ſmells juſt like common air in which the ſame mixture has ſtood ; and it is not capable of being diminiſhed any farther, by a freſh mixture of iron and brimſtone.

Common air ſaturated with nitrous air is alſo no farther diminiſhed by this mixture of iron filings and brimſtone, though the mixture ferments with great heat, and ſwells very much in it.

Plants die very ſoon, both in nitrous air, and alſo in common air ſaturated with nitrous air, but eſpecially in the former.

Neither nitrous air, nor common air ſaturated with nitrous air, differ in ſpecific gravity from common air. At leaſt, the difference is ſo ſmall, that I could not be ſure there was any ; ſometimes about three pints of it ſeeming to be about half a grain heavier, and at other times as much lighter than common air.

I 4   Having

Having, among other kinds of air, expofed a quantity of nitrous air to water out of which the air had been well boiled, in the experiment to which I have more than once referred (as having been the occafion of feveral new and important obfervations) I found that $\frac{12}{10}$ of the whole was abforbed. Perceiving, to my great furprize, that fo very great a proportion of this kind of air was mifcible with water, I immediately began to agitate a confiderable quantity of it, in a jar ftanding in a trough of the fame kind of water; and, with about four times as much agitation as fixed air requires, it was fo far abforbed by the water, that only about one fifth remained. This remainder extinguifhed flame, and was noxious to animals.

Afterwards I diminifhed a pretty large quantity of nitrous air to one eighth of its original bulk, and the remainder ftill retained much of its peculiar fmell, and diminifheu common air a little. A moufe alfo died in it, but not fo fuddenly as it would have done in pure nitrous air. In this operation the peculiar fmell of nitrous air is very manifeft, the water being firft impregnated with the air, and then tranfmitting it to the common atmofphere.

This experiment gave me the hint of impregnating water with nitrous air, in the manner

ner in which I had before done it with fixed air;
and I prefently found that diftilled water would
imbibe about one tenth of its bulk of this kind
of air, and that it acquired a remarkably acid
and aftringent tafte from it. The fmell of
water thus impregnated is at firft peculiarly
pungent. I did not chufe to fwallow any of it,
though, for any thing that I know, it may be
perfectly innocent, and perhaps, in fome cafes,
falutary.

This kind of air is retained very obftinately
by water. In an exhaufted receiver a quantity
of water thus faturated emitted a whitifh fume,
fuch as fometimes iffues from bubbles of this air
when it is firft generated, and alfo fome air-
bubbles; but though it was fuffered to ftand
a long time in this fituation, it ftill retained its
peculiar tafte; but when it had ftood all night
pretty near the fire, the water was become quite
vapid, and had depofited a filmy kind of matter, of which I had often collected a confiderable quantity from the trough in which jars
containing this air had ftood. This I fuppofe
to be a precipitate of the metal, by the folution
of which the nitrous air was generated. I have
not given fo much attention to it as to know,
with certainty, in what circumftances this *depofit* is made, any more than I do the matter
depofited from inflammable air above-mentioned;

I

ed; for I cannot get it, at leaft in any confiderable quantity, when I pleafe; whereas I
have often found abundance of it, when I did
not expect it at all.

The nitrous air with which I made the firft
impregnation of water was extracted from copper; but when I made the impregnation with
air from quickfilver, the water had the very
fame tafte, though the matter depofited from it
feemed to be of a different kind; for it was
whitifh, whereas the other had a yellowifh tinge.
Except the firft quantity of this impregnated
water, I could never deprive any more that I
made of its peculiar tafte. I have even let fome
of it ftand more than a week, in phials with their
mouths open, and fometimes very near the fire,
without producing any alteration in it *.

Whether any of the fpirit of nitre contained
in the nitrous air be mixed with the water in
this operation, I have not yet endeavoured to
determine. This, however, may probably
be the cafe, as the fpirit of nitre is, in a confiderable degree, volatile †.

It

* I have fince found, that nitrous air has never failed
to efcape from the water, which has been impregnated with
it, by long expofure to the open air.
† This fufpicion has been confirmed by the ingenious
Mr. Bewley, of Great Maffingham in Norfolk, who has difcovered

It will perhaps be thought, that the moſt *uſeful,* if not the moſt remarkable, of all the properties of this extraordinary kind of air, is its power of preſerving animal ſubſtances from putrefaction, and of reſtoring thoſe that are already putrid, which it poſſeſſes in a far greater degree than fixed air. My firſt obſervation of this was altogether caſual. Having found nitrous air to ſuffer ſo great a diminution as I have already mentioned by a mixture of iron filings and brimſtone, I was willing to try whether it would be equally diminiſhed by other cauſes of the diminution of common air, eſpecially by putrefaction ; and for this purpoſe I put a dead mouſe into a quantity of it, and placed it near the fire, where the tendency to putrefaction was very great. In this caſe there was a conſiderable diminution, *viz.* from $5\frac{1}{4}$ to $3\frac{1}{4}$; but not ſo great as I had expected, the antiſeptic power of the nitrous air having checked the tendency to putrefaction ; for when, after a week, I took the mouſe out, I

covered that the acid taſte of this water is not the neceſſary conſequence of its impregnation with nitrous air, but is the effect of the *acid vapour,* into which part of this air is reſolved, when it is decompoſed by a mixture with common air. This, it will be ſeen, exactly agrees with my own obſervation on the conſtitution of nitrous air, in the ſecond part of this work. A more particular account of Mr. Bewley's obſervation will be given in the *Appendix.*

perceived,

perceived, to my very great surprize, that it had no offensive smell.

Upon this I took two other mice, one of them just killed, and the other soft and putrid, and put them both into the same jar of nitrous air, standing in the usual temperature of the weather, in the months of July and August of 1772; and after twenty-five days, having observed that there was little or no change in the quantity of the air, I took the mice out; and, examining them, found them both perfectly sweet, even when cut through in several places. That which had been put into the air when just dead was quite firm; and the flesh of the other, which had been putrid and soft, was still soft, but perfectly sweet.

In order to compare the antiseptic power of this kind of air with that of fixed air, I examined a mouse which I had inclosed in a phial full of fixed air, as pure as I could make it, and which I had corked very close; but upon opening this phial in water about a month after, I perceived that a large quantity of putrid effluvium had been generated; for it rushed with violence out of the phial; and the smell that came from it, the moment the cork was taken out, was insufferably offensive. Indeed Dr. Macbride says, that he could only restore very thin pieces

of

of putrid flefh by means of fixed air. Perhaps the antifeptic power of thefe kinds of air may be in proportion to their acidity.

If a little pains were taken with this fubject this remarkable antifeptic power of nitrous air might poffibly be applied to various ufes, per- haps to the prefervation of the more delicate birds, fifhes, fruits, &c. mixing it in different proportions with common or fixed air. Of this property of nitrous air anatomifts may perhaps avail themfelves, as animal fubftances may by this means be preferved in their natural foft ftate ; but how long it will anfwer for this pur- pofe, experience only can fhew.

I calcined lead and tin in the manner hereafter defcribed in a quantity of nitrous air, but with very little fenfible effect ; which rather furprized me ; as, from the refult of the experiment with the iron filings and brimftone, I had expected a very great diminution of the nitrous air by this procefs ; the mixture of iron filings and brimftone, and the calcination of metals, having the fame effect upon common air, both of them diminifhing it in nearly the fame proportion. But though I made the metals *fume* copioufly in nitrous air, there might be no real *calcination*, the phlogifton not being feparated, and the pro- per calcination prevented by there being no *fixed*

air,

air, which is neceffary to the formation of the calx, to unite with it.

Nitrous air is procured from all the proper metals by fpirit of nitre, except lead, and from all the femi-metals that I have tried, except zinc. For this purpofe I have ufed bifmuth and nickel, with fpirit of nitre only, and regulus of antimony and platina, with *aqua regia.*

I got little or no air from lead by fpirit of nitre, and have not yet made any experiments to afcertain the nature of this folution. With zinc I have taken a little pains.

Four penny-weights and feventeen grains of zinc diffolved in fpirit of nitre, to which as much water was added, yielded about twelve ounce meafures of air, which had, in fome degree, the properties of nitrous air, making a flight effervefcence with common air, and diminifhing it about as much as nitrous air, which had been itfelf diminifhed one half by wafhing in water. The fmell of them both was alfo the fame ; fo that I concluded it to be the fame thing, that part of the nitrous air, which is imbibed by water, being retained in this folution.

In order to difcover whether this was the cafe, I made the folution boil in a fand-heat. Some

air

air came from it in this ftate, which feemed to
be the fame thing, with nitrous air diminifhed
about one fixth, or one eighth, by wafhing in
water. When the fluid part was evaporated,
there remained a brown fixed fubftance, which
was obferved by Mr. Hellot, who defcribes it,
Ac. Par. 1735, M. p. 35. A part of this I
threw into a fmall red-hot crucible; and cover-
ing it immediately with a receiver, ftanding in
water, I obferved that very denfe red fumes
rofe from it, and filled the receiver. This
rednefs continued about as long as that which
is occafioned by a mixture of nitrous and common air; the air was alfo confiderably diminifhed within the receiver. This fubftance,
therefore, muft certainly have contained within
it the very fame thing, or principle, on which
the peculiar properties of nitrous air depend.

It is remarkable, however, that though the
air within the receiver was diminifhed about
one fifth by this procefs, it was itfelf as much
affected with a mixture of nitrous air, as common air is, and a candle burned in it very well.
This may perhaps be attributed to fome effect
of the fpirit of nitre, in the compofition of that
brown fubftance.

Nitrous air, I find, will be confiderably diminifhed in its bulk by ftanding a long time
in water, about as much as inflammable air is
dimi-

diminifhed in the fame circumftances. For this purpofe I kept for fome months a quart-bottle full of each of thefe kinds of air; but as different quantities of inflammable air vary very much in this refpect, it is not improbable but that nitrous air may vary alfo.

From one trial that I made, I conclude that nitrous air may be kept in a bladder much better than moft other kinds of air. The air to which I refer was kept about a fortnight in a bladder, through which the peculiar fmell of the nitrous air was very fenfible for feveral days. In a day or two the bladder became red, and was much contracted in its dimen-fions. The air within it had loft very little of its peculiar property of diminifhing com-mon air.

I did not endeavour to afcertain the exact quantity of nitrous air produced from given quantities of all the metals which yield it; but the few obfervations which I did make for this purpofe I fhall recite in this place:
dwt. gr.

dwt.	gr.		
6	o	of filver yielded	$17\frac{1}{2}$ ounce meafures.
5	19	of quickfilver	$4\frac{1}{2}$
1	$2\frac{1}{2}$	of copper	$14\frac{1}{2}$
2	o	of brafs	21
o	20	of iron	16
1	5	of bifmuth	6
o	12	of nickel	4

SECTION

SECTION VII.

Of AIR *infeЕted with the* FUMES *of* BURNING CHARCOAL.

Air infeЕted with the fumes of burning charcoal is well known to be noxious; and the Honourable Mr. Cavendiſh favoured me with an account of ſome experiments of his, in which a quantity of common air was reduced from 180 to 162 ounce meaſures; by paſſing through a red-hot iron tube filled with the duſt of charcoal. This diminution he aſcribed to ſuch a *deſtruЕion* of common air as Dr. Hales imagined to be the conſequence of burning. Mr. Cavendiſh alſo obſerved, that there had been a generation of fixed air in this pro-ceſs; but that it was abſorbed by ſope leys. This experiment I alſo repeated, with a ſmall variation of circumſtances, and with nearly the ſame reſult.

Afterwards, I endeavoured to aſcertain, by what appears to me to be an eaſier and more certain method, in what manner air is affeЕted with the fumes of charcoal, viz. by ſuſpending bits of charcoal within glaſs veſſels, filled to a certain height with water, and ſtanding inverted

K   in

in another veffel of water, while I threw the focus of a burning mirror, or lens, upon them.
In this manner I diminifhed a given quantity
of air one fifth, which is nearly in the fame
proportion with other diminutions of air.

If, inftead of pure water, I ufed *lime-water*
in this procefs, it never failed to become turbid
by the precipitation of the lime, which could
only be occafioned by fixed air, either difcharged
from the charcoal, or depofited by the common
air. At firft I concluded that it came from the
charcoal ; but confidering that it is not probable
that fixed air, confined in any fubftance, can
bear fo great a degree of heat as is neceffary to
make charcoal, without being wholly expelled ;
and that in other diminutions of common
air, by phlogifton only, there appears to be a
depofition of fixed air, I have now no doubt
but that, in this cafe alfo, it is fupplied from
the fame fource.

This opinion is the more probable, from
there being the fame precipitation of lime, in
this procefs, with whatever degree of heat the
charcoal had been made. If, however, the charcoal had not been made with a very confiderable
degree of heat, there never failed to be a permanent addition of inflammable air produced ;
which agrees with what I obferved before, that,

in

in converting dry wood into charcoal, the greateft part is changed into inflammable air

I have fometimes found, that charcoal which
was made with the moft intenfe heat of a fmith's
fire, which vitrified part of a common crucible
in which the charcoal was confined, and which
had been continued above half an hour, did not
diminifh the air in which the focus of a burning
mirror was thrown upon it; a quantity of inflammable air equal to the diminution of the
common air being generated in the procefs:
whereas, at other times, I have not perceived
that there was any generation of inflammable
air, but a fimple diminution of common air,
when the charcoal had been made with a much
lefs degree of heat. This fubjeЕt deferves to be
farther inveftigated.

To make the preceding experiment with ftill
more accuracy, I repeated it in quickfilver;
when I perceived that there was a fmall increafe
of the quantity of air, probably from a generation of inflammable air. Thus it ftood without
any alteration a whole night, and part of the
following day; when lime-water, being admitted to it, it prefently became turbid, and, after
fome time, the whole quantity of air, which
was about four ounce meafures, was diminifhed
one fifth, as before. In this cafe, I carefully
weighed

weighed the piece of charcoal, which was exactly two grains, and could not find that it
was fenfibly diminifhed in weight by the operation.

Air thus diminifhed by the fumes of burning
charcoal not only extinguifhes flame, but is in
the higheft degree noxious to animals ; it makes
no effervefcence with nitrous air, and is incapable of being diminifhed any farther by the fumes
of more charcoal, by a mixture of iron filings
and brimftone, or by any other caufe of the diminution of air that I am acquainted with.

This obfervation, which refpects all other
kinds of diminifhed air, proves that Dr. Hales
was miftaken in his notion of the *abforption* of
air in thofe circumftances in which he obferved
it.   For he fuppofed that the remainder was, in
all cafes, of the fame nature with that which
had been abforbed, and that the operation of the
fame caufe would not have failed to produce a
farther diminution ; whereas all my obfervations
fhew that air, which has once been fully
diminifhed by any caufe whatever, is not only
incapable of any farther diminution, either from
the fame or from any other caufe, but that it
has likewife acquired *new properties*, moft remarkably different from thofe which it had before, and that they are, in a great meafure, the
fame.

fame in all the cafes. Thefe circumftances give reafon to fufpect, that the caufe of diminution is, in reality, the fame in all the cafes. What this caufe is, may, perhaps, appear in the next courfe of obfervations.

S E C T I O N   VIII.

Of the effect of the CALCINATION *of* METALS, *and of the* EFFLUVIA *of* PAINT *made with* WHITE-LEAD *and* OIL, *on* AIR.

Having been led to fufpect, from the experiments which I had made with charcoal, that the diminution of air in that cafe, and perhaps in other cafes alfo, was, in fome way or other the confequence of its having more than its ufual quantity of phlogifton, it occurred to me, that the calcination of metals, which are generally fuppofed to confift of nothing but a metallic earth united to phlogifton, would tend to afcertain the fact, and be a kind of *experimentum crucis* in the cafe.

Accordingly, I fufpended pieces of lead and tin in given quantities of air, in the fame manner as I had before treated the charcoal; and throwing the focus of a burning mirror or lens upon them, fo as to make them fume

K 3   co-

copiouſly. I preſently perceived a diminution of the air.   In the firſt trial that I made,
I reduced four ounce meaſures of air to three,
which is the greateſt diminution of common
air that I had ever obſerved before, and which
I account for, by ſuppoſing that, in other caſes,
there was not only a cauſe of diminution, but
cauſes of addition alſo, either of fixed or inflammable air, or ſome other permanently elaſtic matter, but that the effect of the calcination of metals being ſimply the eſcape of phlogiſton, the cauſe of diminution was alone and
uncontrouled.

The air, which I had thus diminiſhed by
calcination of lead, I transferred into another
clean phial, but found that the calcination of
more lead in it (or at leaſt the attempt to make
a farther calcination) had no farther effect upon it.   This air alſo, like that which had been
infected with the fumes of charcoal, was in the
higheſt degree noxious, made no effervescence
with nitrous air, was no farther diminiſhed by
the mixture of iron filings and brimſtone, and
was not only rendered innoxious, but alſo recovered, in a great meaſure, the other properties of common air, by waſhing in water.

It might be ſuſpected that the noxious quality of air in which *lead* was calcined, might
                                                    be

be owing to fome fumes peculiar to that metal ; but I found no fenfible difference between
the properties of this air, and that in which *tin*
was calcined.

The *water* over which metals are calcined
acquires a yellowifh tinge, and an exceedingly
pungent mell and tafte, pretty much (as near
as I can ·ccollc&, for I did not compare them
together) like that over which brimftone has
been frequently burned. Alfo a thin and whitifh pellicle covered both the furface of the
water, and likewife the fides of the phial in
which the calcination was made; infomuch
that, without frequently agitating the water, it
grew fo opaque by this conftantly accumulating
incruftation, that the fun-beams could not be
tranfmitted through it in a quantity fufficient to
produce the calcination.

I imagined, however, that, even when this
air was transferred into a clean phial, the metals were not fo eafily melted or calcined as they
were in frefh air ; for the air being once fully
faturated with phlogifton, may not fo readily
admit any more, though it be only to tranfmit
it to the water. I alfo fufpected that metals
were not eafily melted or calcined in inflammable, fixed, or nitrous air, or any kind of

K 4   dimi-

diminifhed air.* None of thefe kinds of air
fuffered any change by this operation ; nor was
there any precipitation of lime, when charcoal
was heated in any of thefe kinds of air ftanding in lime-water. This furnifhes another, and
I think a pretty decifive proof, that, in the
precipitation of lime by charcoal, the fixed air
does not come from the charcoal, but from
the common air. Otherwife it is hard to affign
a reafon, why the fame degree of heat (or at
leaft a much greater) fhould not expel the fixed
air from this fubftance, though furrounded by
thefe different kinds of air, and why the fixed
air might not be tranfmitted through them to
the lime-water.

Query. May not water impregnated with
phlogifton from calcined metals, or by any
other method, be of fome ufe in medicine ?
The effect of this impregnation is exceedingly
remarkable ; but the principle with which it is
impregnated is volatile, and intirely efcapes in
a day or two, if the furface of the water be
expofed to the common atmofphere.

* I conclude from the experiments of M. Lavoifier,
which were made with a much better burning lens than I
had an opportunity of making ufe of, that there was no
real calcination of the metals, though they were made to
fume in inflammable or nitrous air; becaufe he was not
able to produce more than a flight degree of calcination
in any given quantity of common air.

It

It mould feem that phlogifton is retained more obftinately by charcoal than it is by lead or tin ; for when any given quantity of air is fully faturated with phlogifton from charcoal, no heat that I have yet applied has been able to produce any more effect upon it ; whereas, in the fame circumftances, lead and tin may ftill be calcined, at leaft be made to emit a copious fume, in which fome part of the phlogifton may be fet loofe. The air indeed, can take no more ; but the water receives it, and the fides of the phial alfo receive an addition of incruftation. This is a white powdery fubftance, and well deferves to be examined. I fhall endeavour to do it at my leifure.

Lime-water never became turbid by the calcination of metals over it, the calx immediately feizing the precipitated fixed air, in preference to the lime in the water ; but the colour, fmell, and tafte of the water was always changed and the furface of it became covered with a yellow pellicle, as before.

When this procefs was made in quickfilver, the air was diminifhed only one fifth ; and upon water being admitted to it, no more was abforbed ; which is an effect fimilar to that of a mixture of nitrous and common air, which was mentioned before.

The

The preceding experiments on the calcination of metals fuggefted to me a method of explaining the caufe of the mifchief which is known to arife from frefh *paint*, made with white-lead (which I fuppofe is an imperfect calx of lead) and oil.

To verify my hypothefis, I firft put a fmall pot full of this kind of paint, and afterwards (which anfwered much better, by expofing a greater furface of the paint) I daubed feveral pieces of paper with it, and put them under a receiver, and obferved, that in about twenty-four hours, the air was diminifhed between one fifth and one fourth, for I did not meafure it very exactly. This air alfo was, as I expected to find, in the higheft degree noxious ; it did not effervefce with nitrous air, it was no farther diminifhed by a mixture of iron filings and brimftone, and was made wholefome by agitation in water deprived of all air.

I think it appears pretty evident, from the preceding experiments on the calcination of metals that air is, fome way or other, diminifhed in confequence of being highly charged with phlogifton ; and that agitation in water reftores it, by imbibing a great part of the phlogiftic matter.

That

That water has a confiderable affinity with phlogifton, is evident from the ftrong impregnation which it receives from it. May not plants alfo reftore air diminifhed by putrefaction by abforbing part of the phlogifton with which it is loaded? The greater part of a dry plant, as well as of a dry animal fubftance, confifts of inflammable air, or fomething that is capable of being converted into inflammable air; and it feems to be as probable that this phlogiftic matter may have been imbibed by the roots and leaves of plants, and afterwards incorporated into their fubftance, as that it is altogether produced by the power of vegetation. May not this phlogiftic matter be even the moft effential part of the food and fupport of both vegetable and animal bodies?

In the experiments with metals, the diminution of air feems to be the confequence of nothing but a faturation with phlogifton; and in all the other cafes of the diminution of air, I do not fee but that it may be effected by the fame means. When a vegetable or animal fubftance is diffolved by putrefaction, the efcape of the phlogiftic matter (which, together with all its other conftituent parts, is then let loofe from it) may be the circumftance that produces the diminution of the air in which it putrefies. It is highly improbable that what remains after an
ani-

animal body has been thoroughly diffolved by putrefaction, fhould yield fo great a quantity of inflammable air, as the dried animal fub-ftance would have done. Of this I have not made an actual trial, though I have often thought of doing it, and ftill intend to do it; but I think there can be no doubt of the refult.

Again, iron, by its fermentation with brim-ftone and water, is evidently reduced to a calx, fo that phlogifton muft have efcaped from it. Phlogifton alfo muft evidently be fet loofe by the ignition of charcoal, and is not improbably the matter which flies off from paint, compofed of white-lead and oil. Laftly, fince fpirit of nitre is known to have a very remarkable affini-ty with phlogifton, it is far from being impro-bable that nitrous air may alfo produce the fame effect by the fame means.

To this hypothefis it may be objected, that, if diminifhed air be air faturated with phlo-gifton, it ought to be inflammable. But this by no means follows; fince its inflammability may depend upon fome particular *mode of com-bination*, or degree of affinity, with which we are not acquainted. Befides, inflammable air feems to confift of fome other principle, or to have fome other conftituent part, befides phlo-gifton

gifton and common air, as is probable from
that remarkable depofit, which, as I have obferved, is made by inflammable air, both from
iron and zinc.

It is not improbable, however, but that a
gréater degree of heat may inflame that air
which extinguifhes a common candle, if it
could be conveniently applied. Air that is inflammable, I obferve, extinguifhes red-hot
wood; and indeed inflammable fubftances can
only be thofe which, in a certain degree of
heat, have a lefs affinity with the phlogifton
they contain, than the air, or fome other contiguous fubftance, has with it; fo that the
phlogifton only quits one fubftance, with which
it was before combined, and enters another,
with which it may be combined in a very different manner. This fubftance, however, whether it be air or any thing elfe, being now
fully faturated with phlogifton, and not being
able to take any more, in the fame circumftances, muft neceffarily extinguifh fire, and
put a ftop to the ignition of all other bodies,
that is, to the farther efcape of phlogifton from
them.

That plants reftore noxious air, by imbibing
the phlogifton with which it is loaded, is very
agreeable to the conjectures of Dr. Franklin,
made

made many years ago, and expreffed in the following extract from the laft edition of his Letters, p. 346.

" I have been inclined to think that the fluid
" *fire*, as well as the fluid *air*, is attracted by
" plants in their growth, and becomes con-
" folidated with the other materials of which
" they are formed, and makes a great part of
" their fubftance ; that, when they come to
" be digefted, and to fuffer in the veffels a
" kind of fermentation, part of the fire, as
" well as part of the air, recovers its fluid
" active ftate again, and diffufes itfelt in the
" body, digefting and feparating it ; that the
" fire fo re-produced, by digeftion and fepara-
" tion, continually leaving the body, its place
" is fupplied by frefh quantities, arifing from
" the continual feparation ; that whatever
" quickens the motion of the fluids in an
" animal, quickens the feparation, and re-pro-
" duces more of the fire, as exercife ; that all
" the fire emitted by wood, and other com-
" buftibles, when burning, exifted in them be-
" fore in a folid ftate, being only difcovered
" when feparating ; that fome foffils, as ful-
" phur, fea-coal, &c. contain a great deal of
" folid fire ; and that, in fhort, what efcapes
" and is diffipated in the burning of bodies,
" befides water and earth, is generally the
I   air

" air and fire, that before made parts of the
" folid."

S E C T I O N IX.

Of MARINE ACID AIR.

Being very much ftruck with the refult of
an experiment of the Hon. Mr. Cavendifh, related Phil. Tranf. Vol. LVI. p. 157, by which,
though, he fays, he was not able to get any
inflammable air from copper, by means of fpirit of falt, he got a much more remarkable
kind of air, *viz.* one that loft its elafticity by
coming into contact with water, I was exceedingly defirous of making myfelf acquainted
with it. On this account, I began with making the experiment in quickfilver, which I
never failed to do in any cafe in which I fufpected that air might either be abforbed by
water, or be in any other manner affected by
it; and by this means I prefently got a much
more diftinct idea of the nature and effects of
this curious folution.

Having put fome copper filings into a fmall
phial, with a quantity of fpirit of falt; and
making the air (which was generated in great
plenty, on the application of heat) afcend into
a tall

a tall glaſs veſſel full of quickſilver, and ſtand
ing in quickſilver, the whole produce continued a conſiderable time without any change
of dimenſions. I then introduced a ſmall
quantity of water to it; when about three
fourths of it (the whole being about four
ounce meaſures) preſently, but gradually, diſappeared, the quickſilver riſing in the veſſel.
I then introduced a conſiderable quantity of
water ; but there was no farther diminution of
the air, and the remainder I found to be inflammable.

Having frequently continued this proceſs a
long time after the admiſſion of the water, I
was much amuſed with obſerving the large bubbles of the newly generated air, which came
through the quickſilver, the ſudden diminution
of them when they came to the water, and the
very ſmall bubbles which went through the water. They made, however, a continual, though
ſlow, increaſe of inflammable air.

Fixed air, being admitted to the whole produce of this air from copper, had no ſenſible
effect upon it. Upon the admiſſion of water,
a great part of the mixture preſently diſappeared; another part, which I ſuppoſe to have
been the fixed air, was abſorbed ſlowly ; and
in this particular caſe the very ſmall permanent
                                              reſi-

refiduum did not take fire; but it is very poffible that it might have done fo, if the quantity
had been greater.

The folution of *lead* in the marine acid is attended with the very fame phænomena as the
folution of copper in the fame acid; about
three fourths of the generated air difappearing
on the admiffion of water; and the remainder
being inflammable.

The folutions of iron, tin, and zinc, in the
marine acid, were all attended with the fame
phænomena as the folutions of copper and
lead, but in a lefs degree; for in iron one
eighth, in tin one fixth, and in zinc one tenth
of the generated air difappeared on the admiffion
with water. The remainder of the air from
iron, in this cafe, burned with a green, or very
light blue flame.

I had always thought it fomething extraordinary that a fpecies of air fhould *lofe its elafticity* by the mere *contact* of any thing, and from
the firft fufpected that it muft have been *imbibed* by the water that was admitted to it; but
fo very great a quantity of this air difappeared upon the admiffion of a very fmall quantity of water, that at firft I could not help concluding that appearances favoured the former
L   hy-

hypothefis. I found, however, that when I
admitted a much ſmaller quantity of water,
confined in a narrow glaſs tube, a part only of
the air diſappeared, and that very ſlowly, and
that more of it vaniſhed upon the admiſſion of
more water. This obſervation put it beyond
a doubt, that this air was properly *imbibed*
by the water, which, being once fully ſaturated with it, was not capable of receiving any
more.

The water thus impregnated taſted very acid,
even when it was much diluted with other water, through which the tube containing it was
drawn. It even diſſolved iron very faſt, and
generated inflammable air. This laſt obſervation, together with another which immediately
follows, led me to the diſcovery of the true nature of this remarkable kind of air.

Happening, at one time, to uſe a good deal
of copper and a ſmall quantity of ſpirit of
ſalt, in the generation of this kind of air, I
was ſurprized to find that air was produced
long after, I could not but think that the acid
muſt have been ſaturated with the metal ; and
I alſo found that the proportion of inflammable air to that which was abſorbed by the water
continually diminiſhed, till, inſtead of being
one

one fourth of the whole, as I had firſt obſerved,
it was not ſo much as one twentieth. Upon
this, I concluded that this ſubtle air did not
ariſe from the copper, but from the ſpirit of
ſalt; and preſently making the experiment with
the acid only, without any copper, or metal of
any kind, this air was immediately produced
in as great plenty as before; ſo that this remarkable kind of air is, in faċt, nothing more
than the vapour, or fumes of ſpirit of ſalt,
which appear to be of ſuch a nature, that they
are not liable to be condenſed by cold, like the
vapour of water, and other fluids, and therefore may be very properly called an *acid air*,
or more reſtriċtively, the *marine acid air*.

This elaſtic acid vapour, or acid air, extinguiſhes flame, and is much heavier than common air; but how much heavier, will not be
eaſy to aſcertain. A cylindrical glaſs veſſel,
about three fourths of an inch in diameter,
and four inches deep, being filled with it, and
turned upſide down, a lighted candle may be
let down into it more than' twenty times before
it will burn at the bottom. It is pleaſing to
obſerve the colour of the flame in this experiment; for both before the candle goes out, and
alſo when it is firſt lighted again, it burns with
a beautiful green, or rather light-blue flame,
ſuch as is ſeen when common ſalt is thrown into the fire.　　　L 2　　　When

When this air is all expelled from any quantity of fpirit of falt, which is eafily perceived
by the fubfequent vapour being condenfed by
cold, the remainder is a very weak acid, barely
capable of diffolving iron.

Being now in the poffeffion of a new fubject
of experiments, *viz.* an elaftic acid vapour, in
the form of a permanent air, eafily procured,
and effectually confined by glafs and quickfilver, with which it did not feem to have any
affinity ; I immediately began to introduce a
variety of fubftances to it, in order to afcertain its peculiar properties and affinities, and
alfo the properties of thofe other bodies with
refpect to it.

Beginning with *water*, which, from preceding obfervations, I knew would imbibe it, and
become impregnated with it ; I found that $2\frac{1}{2}$
grains of rain-water abforbed three ounce meafures of this air, after which it was increafed
one third in its bulk, and weighed twice as
much as before ; fo that thihs concentrated vapour feems to be twice as heavy as rain-water :
Water impregnated with it makes the ftrongeft
fpirit of falt that I have feen, diffolving iron
with the moft rapidity. Confequently, two
thirds of the beft fpirit of falt is nothing more
than mere phlegm or water.

Iron

Iron filings, being admitted to this air, were diffolved by it pretty faft, half of the air difappearing, and the other half becoming inflammable air, not abforbed by water. Putting chalk to it, fixed air was produced.

I had not introduced many fubftances to this air, before I difcovered that it had an affinity with *phlogifton*, fo that it would deprive other fubftances of it, and form with it fuch an union as conftitutes inflammable air; which feems to fhew, that inflammable air univerfally confifts of the union of fome acid vapour with phlogifton.

Inflammable air was produced, when to this acid air I put fpirit of wine, oil of olives, oil of turpentine, charcoal, phofphorus, bees-wax, and even fulphur. This laft obfervation, I own, furprized me; for, the marine acid being reckoned the weakeft of the three mineral acids, I did not think that it had been capable of diflodging the oil of vitriol from this fubftance; but I found that it had the very fame effect both upon alum and nitre; the vitriolic acid in the former cafe, and the nitrous in the latter, giving place to the ftronger vapour of fpirit of falt.

L 3

The

The ruft of iron, and the precipitate of 'nitrous air made from copper, alfo imbibed this air very faft, and the little that remained of it was inflammable air; which proves, that thefe calces contain phlogifton. It feems alfo to be pretty evident, from this experiment, that the precipitate above mentioned is a real calx of the metal, by the folution of which the nitrous air is generated.

As fome remarkable circumftances attend the abforption of this acid air, by the fubftances above-mentioned, I fhall briefly mention them.

Spirit of wine abforbs this air as readily as water itfelf, and is increafed in bulk by that means. Alfo, when it is faturated, it diffolves iron with as much rapidity, and ftill continues inflammable.

Oil of olives abforbs this air very flowly, and at the fame time, it turns almoft black, and becomes glutinous. It is alfo lefs mifcible with water, and acquires a very difagreeable fmell. By continuing upon the furface of the water, it became white, and its offenfive fmell went off in a few days.

Oil of turpentine abforbed this air very faft, turning brown, and almoft black. No inflammable

mable air was formed, till I raifed more of the acid air than the oil was able to abforb, and let it ftand a confiderable time ; and ftill the air was but weakly inflammable. The fame was the cafe with the oil of olives, in the laft mentioned experiment ; and it feems to be probable, that, the longer this acid air had continued in contact with the oil, the more phlogifton it would have extracted from it. It is not wholly improbable, but that, in the intermediate ftate, before it becomes inflammable air, it may be nearly of the nature of common air.

Bees-wax abforbed this air very flowly. About the bignefs of a hazel-nut of the wax being put to three ounce meafures of the acid air, the air was diminifhed one half in two days, and, upon the admiffion of water, half of the remainder alfo difappeared. This air was ftrongly inflammable.

Charcoal abforbed this air very faft. About one fourth of it was rendered immifcible in water, and was but weakly inflammable.

A fmall bit of *phofphorus*, perhaps about half a grain, fmoked, and gave light in the acid air, juft as it would have done in common air confined. It was not fenfibly wafted after-

L 4   con-

continuing about twelve hours in that ſtate, and the bulk of the air was very little dimi-niſhed. Water being admitted to it abſorbed it as before, except about one fifth of the whole, It was but weakly inflammable.

Putting ſeveral pieces of *ſulphur* to this air, it was abſorbed but ſlowly. In about twenty-four hours about one fifth of the quantity had diſappeared ; and water being admitted to the remainder, very little more was abſorbed. The remainder was inflammable, and burned with a blue flame.

Notwithſtanding the affinity which this acid air appears to have with phlogiſton, it is not capable of depriving all bodies of it. I found that dry wood, cruſts of bread, and raw fleſh, very readily imbibed this air, but did not part with any of their phlogiſton to it. All theſe ſubſtances turned very brown, after they had been ſome time expoſed to this air, and taſted very ſtrongly of the acid when they were taken out ; but the fleſh, when waſhed in water, be-came very white, and the fibres eaſily ſeparated from one another, even more than they would have done if it had been boiled or roaſted *

* It will be ſeen, in the ſecond part of this work, that, in ſome of theſe proceſſes, I had afterwards more ſucceſs.

When

When I put a piece of *faltpetre* to that is it was prefently furrounded with a white fume, which foon filled the whole veffel, exactly like the fume which burfts from the bubbles of nitrous air, when it is generated by a vigorous fermentation, and fuch as is feen when nitrous air is mixed with this acid air. In about a minute, the whole quantity of air was abforbed, except a very little, which might be the common air that had lodged upon the furface of the fpirit of falt within the phial.

A piece of *alum* expofed to this air turned yellow, abforbed it as faft as the faltpetre had done, and was reduced by it to the form of a powder. Common falt, as might be expected, had no effect whatever on this marine acid air.

I had alfo imagined, that if air diminifhed by the proceffes above-mentioned was affected in this manner, in confequence of its being faturated with phlogifton, a mixture of this acid air might imbibe that phlogifton, and render it wholefome again ; but I put about one fourth of this air to a quantity of air in which metals had been calcined, without making any fenfible alteration in it. I do not, however, infer from this, that air is not diminifhed by means of phlogifton, fince the common air, like fome other fubftances, may hold
the

the phlogifton too faft, to be -deprived of it by this acid air.

I fhall conclude my account of thefe experiments with obferving, that the electric fpark is vifible in acid air, exactly as it is in common air; and though I kept making this fpark a confiderable time in a quantity of it, I did not perceive that any fenfible alteration was made in it. A little inflammable air was produced, but not more than might have come from the two iron nails which I made ufe of in taking the fparks.

SECTION X.

MISCELLANEOUS OBSERVATIONS.

1. As many of the preceding obfervations relate to the *vinous* and *putrefactive* fermentations, I had the curiofity to endeavour to afcertain in what manner the air would be affected by the *acetous* fermentation. For this purpofe I inclofed a phial full of fmall beer in a jar ftanding in water; and obferved that, during the firft two or three days, there was an increafe of the air in the jar, but from that time it gradually decreafed, till at length there appeared to be a diminution of about one tenth of the whole quantity.

During

During this time the whole furface of it was gradually covered with a fcum, beautifully corrugated. After this there was an increafe of the air till there was more than the original quantity ; but this muft have been fixed air, not incorporated with the reft of the mafs ; for, withdrawing the beer, which I found to be four, after it had ftood 18 or 20 days under the jar, and paffing the air feveral times through cold water, the original quantity was diminifhed about one ninth. In the remainder a candle would not burn, and a moufe would have died prefently.

The fmell of this air was exceedingly pungent, but different from that of the putrid effluvium. A moufe lived perfectly well in this air, thus affected with the acetous fermentation ; after it had ftood feveral days mixed with four times the quantity of fixed air.

2. All the kinds of factitious air on which I have yet made the experiment are highly noxious, except that which is extracted from faltpetre, or alum ; but in this even a candle burned juft as in common air*. In one quantity which
I got

---

* Experiments, of which an account will be given in the fecond part of this work, make it probable, that though a candle burned even *more than well* in this air, an
animal

I got from faltpetre a candle not only burned, but the flame was increafed, and fomething was heard like a hiffing, fimilar to the decrepitation of nitre in an open fire. This experiment was made when the air was frefh made, and while it probably contained fome particles of nitre, which would have been depofited afterwards. The air was extraƐted from thefe fubftances by heating them in a gun-barrel, which was much corroded and foon fpoiled by the experiment. What effeƐt this circumftance may have had upon the air I have not confidered.

November 6, 1772, I had the curiofity to examine the ftate of a quantity of this air which had been extraƐted from faltpetre above a year, and which·at firft was perfeƐtly wholefome ; when, to my very great furprize, I found that it was become, in the higheft degree, noxious. It made no effervefcence with nitrous air, and a moufe died the moment it was put into it. I had not, however, wafhed it in rainwater quite ten minutes (and perhaps lefs time would have been fufficient) when I found, upon trial, that it was reftored to its former perfeƐtly wholefome ftate. It effervefced with nitrous air as much as the beft common air ever does, and

animal would not have lived in it. At the time of this firft publication, however, I had no idea of this being poffible in nature.

even

even a candle burned in it very well, which I had never before obferved of any kind of noxious air meliorated by agitation in water. This feries of facts, relating to air extracted from nitre, appear to me to be very extraordinary and important, and, in able hands, may lead to confiderable difcoveries.

3. There are many fubftances which impregnate common air in a very remarkable manner, but without making it noxious to animals. Among other things I tried volatile alkaline falts, and camphor; the latter of which I melted with a burning-glafs, in air inclofed in a phial. The moufe, which was put into this air, fneezed and coughed very much, efpecially after it was taken out; but it prefently recovered, and did not appear to have been fenfibly injured.

4. Having made feveral experiments with a mixture of iron filings and brimftone, kneaded to a pafte with water, I had the curiofity to try what would be the effect of fubftituting *brafs duft* in the place of the iron filings. The refult was, that when this mixture had ftood about three weeks, in a given quantity of air, it had turned black, but was not increafed in bulk. The air alfo was neither fenfibly increafed nor decreafed, but the nature of it was changed; for it extinguifhed flame, it would have
killed

killed a mouse prefently, and was not reftored by fixed air, which had been mixed
with it feveral days.

5. I have frequently mentioned my having,
at one time, expofed equal quantities of different kinds of air in jars ftanding in boiled water.
Common air in this experiment was diminifhed
four fevenths, and the remainder extinguifhed
flame. This experiment demonftrates that water
does not abforb air equally, but that it decom-
pofes it, taking one part, and leaving the reft.
To be quite fure of this fact, I agitated a
quantity of common air in boiled water, and
when I had reduced it from eleven ounce meafures to feven, I found that it extinguifhed a
candle, but a moufe lived in it very well. At
another time a candle barely went out when the
air was diminifhed one third, and at other times
I have found this effect take place at other very
different degrees of diminution.

This difference I attribute to the differences
in the ftate of the water with refpect to the air
contained in it; for fometimes it had ftood
longer than at other times before I made ufe of
it. I alfo ufed diftilled-water, rain-water, and
water out of which the air had been pumped,
promifcuoufly with rain water. I even doubt,
not but that, in a certain ftate of the water,
there

there might be no fenfible difference in the bulk of the agitated air, and yet at the end of the procefs it would extinguifh a candle, air being fupplied from the water in the place of that part of the common air which had been ab-forbed.

It is certainly a little extraordinary that the very fame procefs fhould fo far mend putrid air, as to reduce it to the ftandard of air in which candles have burned out ; and yet that it fhould fo far injure common and wholefome air as to reduce it to about the fame ftandard : but fo the fact certainly is. If air extinguifh flame in confequence of its being previoufly faturated with phlogifton, it muft, in this cafe, have been transferred from the water to the air , and it is by no means inconfiftent with this hypothefis to fuppofe, that, if the air be over faturated with phlogifton, the water will imbibe it, till it be reduced to the fame proportion that agitation in water would have communicated to it.

To a quantity of common air, thus dimi-nifhed by agitation in water, till it extinguifhed a candle, I put a plant, but it did not fo far re-ftore it as that a candle would burn in it again ;· which to me appeared not a little extraordinary, as it did not feem to be in a worfe ftate than air in which candles had burned out, and which

which had never failed to be reftored by the fame means.

I had no better fuccefs with a quantity of permanent air which I had collected from my pump-water. Indeed thefe experiments were begun before I was acquainted with that pro- perty of nitrous air, which makes it fo accu- rate a meafure of the goodnefs of other kinds of air ; and it might perhaps be rather too late in the year when I made the experiments. Having neglected thefe two jars of air, the plants died and putrefied in both of them ; and then I found the air in them both to be highly noxious, and to make no effervefcence with nitrous air.

I found that a pint of my pump - water contained about one fourth of an ounce meafure of air, one half of which was afterwards ab- forbed by ftanding in frefh pump-water. A candle would not burn in this air, but a moufe lived in it very well. Upon the whole, it feemed to be in about the fame ftate as air in which a candle had burned out.

6. I once imagined that, by mere *ftagnation*, air might become unfit for refpiration, or at leaft the

the burning of candles ; but if this be the caſe, and the change be produced gradually, it muſt require a long time for the purpoſe. For on the 22d of September 1772, I examined a quantity of common air, which had been kept in a phial, without agitation, from May 1771, and found it to be in no reſpect worſe than freſh air, even by the teſt of the nitrous air.

7. The cryſtallization of nitre makes no ſenſible alteration in the air in which the proceſs is made. For this purpoſe I diſſolved as much nitre as a quantity of hot water would contain, and let it cool under a receiver, ſtanding in water.

8. November 6, 1772, a quantity of inflam mable air, which, by long keeping, had come to extinguiſh flame, I obſerved to ſmell very much like common air in which a mixture of iron filings and brimſtone had ſtood. It was not, however, quite ſo ſtrong, but it was equally noxious.

9. Biſmuth and nickel are diſſolved in the marine acid with the application of a conſiderable degree of heat; but little or no air is got from either of them; but, what I thought a little re-

M   mark-

markable, both of them smelled very much like Harrowgate water, or liver of sulphur. This smell I have met with several times in the course of my experiments, and in processes very different from one another.

PART

PART II.

Experiments and Obfervations made in the Year 1773, and the Beginning of 1774.

SECTION I.

Obfervations on ALKALINE AIR.

AFTER I had made the difcovery of the *marine acid air*, which the vapour of fpirit of falt may properly enough be called, and had made thofe experiments upon it, of which I have given an account in the former part of this work, and others which I propofe to recite in this part; it occurred to me, that, by a procefs fimilar to that by which this *acid* air is expelled from the fpirit of falt, an *alkaline* air might be expelled from fubftances containing volatile alkali.

Ac.

Accordingly I procured ſome volatile ſpirit of
ſal ammoniac, and having put it into a thin
phial, and heated it with the flame of a candle,
I preſently found that a great quantity of vapour was diſcharged from it; and being received in a veſſel of quickſilver, ſtanding in a
baſon of quickſilver, it continued in the form
of a tranſparent and permanent air, not at all
condenſed by cold; ſo that I had the ſame opportunity of making experiments upon it, as I
had before on the acid air, being in the ſame
favourable circumſtances.

With the ſame eaſe I alſo procured this air
from *ſpirit of hartſhorn*, and *ſal volatile* either
in a fluid or ſolid form, i. e. from thoſe volatile
alkaline ſalts which are produced by the diſtillation of ſal ammoniac with fixed alkalis.
But in this caſe I ſoon found that the alkaline
air I procured was not pure; for the fixed air,
which entered into the compoſition of my materials, was expelled along with it. Alſo, uniting again with the alkaline air, in the glaſs
tube through which they were conveyed, they
ſtopped it up, and were often the means of
burſting my veſſels.

While theſe experiments were new to me, I
imagined that I was able to procure this air
with peculiar advantage and in the greateſt
                                                    abun-

abundance, either from the falts in a dry ftate, when they were juft covered with water, or in a perfectly fluid ftate; for, upon applying a candle to the phials in which they were contained, there was a moft aftonifhing production of air; but having examined it, I found it to be chiefly fixed air, efpecially after the firft or fecond produce from the fame materials; and removing my apparatus to a trough of water and ufing the water inftead of quickfilver, I found that it was not prefently abforbed by it.

This, however, appears to be an eafy and elegant method of procuring fixed air, from a fmall quantity of materials, though there muft be a mixture of alkaline air along with it; as it is by means of its combination with this principle only, that it is poffible, that fo much fixed air fhould be retained in any liquid. Water, at leaft, we know, cannot be made to contain much more than its own bulk of fixed air.

After this difappointment, I confined myfelf to the ufe of that volatile fpirit of fal ammoniac which is procured by a diftillation with flaked lime, which contains no fixed air; and which feems, in a general ftate, to contain about as much alkaline air, as an equal quantity of fpirit of falt contains of the acid air.

M 3   Wanting

Wanting, however, to procure this air in greater quantities, and this method being rather expenſive, it occurred to me, that alkaline air might, probably, be procured, with the moſt eaſe and convenience, from the original materials, mixed in the ſame proportions that chemiſts had found by experience to anſwer the beſt for the production of the volatile ſpirit of ſal ammoniac. Accordingly I mixed one fourth of pounded ſal ammoniac, with three fourths of ſlaked lime ; and filling a phial with the mixture, I preſently found it completely anſwered my purpoſe. The heat of a candle expelled from this mixture a prodigious quantity of alkaline air ; and the ſame materials (as much as filled an ounce phial) would ſerve me a conſiderable time, without changing ; eſpecially when, inſtead of a glaſs phial, I made uſe of a ſmall iron tube, which I find much more convenient for the purpoſe.

As water ſoon begins to riſe in this proceſs, it is neceſſary, if the air is intended to be conveyed perfectly *dry* into the veſſel of quickſilver, to have a ſmall veſſel in which this water (which is the common volatile ſpirit of ſal ammoniac) may be received. This ſmall veſſel muſt be interpoſed between the veſſel which contains the materials for the generation of the air, and that in which it is to be received, as *d* fig. 8.

This

This *alkaline* air being perfectly analogous to the *acid* air, I was naturally led to inveſtigate the properties of it in the ſame manner, and nearly in the ſame order. From this analogy I concluded as I preſently found to be the fact, that this alkaline air would be readily imbibed by water, and, by its union with it, would form a volatile ſpirit of ſal ammoniac. And as the water, when admitted to the air in this manner, confined by quickſilver, has an opportunity of fully ſaturating itſelf with the alkaline vapour, it is made prodigiouſly ſtronger than any volatile ſpirit of ſal ammoniac that I have ever ſeen ; and I believe ſtronger than it can be made in the common way.

In order to aſcertain what addition, with reſpect to quantity and weight, water would acquire by being ſaturated with alkaline air, I put 1 ¼ grains of rain-water into a ſmall glaſs tube, cloſed at one end with cement, and open at the other, the column of water meaſuring $\frac{7}{10}$ of an inch ; and having introduced it through the quickſilver into a veſſel containing alkaline air, obſerved that it abſorbed ¼ of an ounce meaſure of the *air*, ·and had then gained about half a grain in weight, and was increaſed to 8 ½ tenths of an inch in length. I did not make a ſecond experiment of this kind, and ·therefore will not anſwer for the exactneſs of theſe proportions in

future

future trials. What I did fufficiently anfwered my purpofe, in a general view of the fub-ject.

When I had, at one time, faturated a quan-tity of diftilled water with alkaline air, fo that a good deal of the air remained unabforbed on the furface of the water, I obferved that, as I continued to throw up more air, a confiderable proportion of it was imbibed, but not the whole; and when I had let the apparatus ftand a day, much more of the air that lay on the furface was imbibed. And after the water would imbibe no more of the *old* air, it imbibed *new*. This fhews that water requires a confiderable time to faturate itfelf with this kind of air, and that part of it more readily unites with water than the reft.

The fame is alfo, probably, the cafe with all the kinds of air with which water can be im-pregnated. Mr. Cavendifh made this obferva-tion with refpect to fixed air, and I repeated the whole procefs above-mentioned with acid air, and had precifely the fame refult. The al-kaline water which I procured in this experi-ment was, beyond comparifon, ftronger to the fmell, than any fpirit of fal ammoniac that I had feen.

This

This experiment led me to attempt the making of fpirit of fal ammoniac in a larger quantity, by impregnating diftilled water with this
alkaline air For this purpofe I filled a piece of
a gun-barrel with the materials above-mentioned,
and luted to the o en end of it a fmall glafs tube,
one end of which was bent, and put within the
mouth of a glafs veffel, containing a quantity of
diftilled water upon quickfilver, ftanding in a
bafon of quickfilver, as in fig. 7. In thefe circumftances the heat of the fire, applied gradually, expelled the alkaline air, which, paffing through
the tube, and the quickfilver, came at laft to the
water, which, in time, became fully faturated
with it.

By this means I got a very ftrong alkaline
liquor, from which I could again expel the alkaline air which I had put into it, whenever it
happened to be more convenient to me to get it
in that manner. This procefs may eafily be
performed in a ftill larger way ; and by this
means a liquor of the fame nature with the volatile fpirit of fal ammoniac, might be made
much ftronger, and much cheaper, than it is
now made.

Having fatisfied myfelf with refpect to the
relation that alkaline air bears to water, I was
impatient to find what would be the confequence
of

of mixing this new air with the other kinds
with which I was acquainted before, and efpecially with *acid* air ; having a notion that thefe
two airs, being of oppofite natures, might com-
pofe a *neutral air*, and perhaps the very fame
thing with common air. But the moment that
thefe two kinds of air came into contact, a beautiful white cloud was formed, and prefently
filled the whole veffel in which they were contained. At the fame time the quantity of air
began to diminifh, and, at length, when the
cloud was fubfided, there appeared to be formed
a folid *white falt*, which was found to be the
common *fal ammoniac*, or the marine acid united
to the volatile alkali.

The firft quantity that I produced immediately deliquefced, upon being expofed to the
common air ; but if it was expofed in a very
dry and warm place, it almoft all evaporated,
in a white cloud. I have, however, fince, from
the fame materials, produced the falt abovementioned in a ftate not fubject to deliquefce or
evaporate. This difference, I find, is owing
to the proportion of the two kinds of air in the
compound. It is only volatile when there is
more than a due proportion of either of the
conftituent parts. In thefe cafes the fmell of the
falts is extremely pungent, but very different
from one another ; being manifeftly acid, or alkaline,

kaline, according to the prevalence of each of
thefe airs refpectively.

Nitrous air admitted to alkaline air likewife
occafioned a whitifh cloud, and part of the air
was abforbed; but it prefently grew clear again;
leaving only a little dimnefs on the fides of the
veffel. This, however, might be a kind of
falt, formed by the union of the two kinds of
air. There was no other falt formed that I
could perceive. Water being admitted to this
mixture of nitrous and alkaline air prefently abforbed the latter, and left the former poffeffed
of its peculiar properties.

Fixed air admitted to alkaline air formed oblong and flender cryftals, which croffed one
another, and covered the fides of the veffel in
the form of net-work. Thefe cryftals muft
be the fame thing with the volatile alkalis
which chemifts get in a folid form, by the diftillation of fal ammoniac with fixed alkaline
falts.

Inflammable air admitted to alkaline air exhibited no particular appearance. Water, as
in the former experiment, abforbed the alkaline air, and left the inflammable air as it was
before. It was remarkable, however, that the
water which was admitted to them became
whitifh,

whitifh, and that this white cloud fettled, in the form of a white powder, to the bottom of the veffel.

Alkaline air mixed with *common air*, and ftanding together feveral days, firft in quick-filver, and then in water (which abforbed the alkaline air) it did not appear that there was any change produced in the common air : at leaft it was as much diminifhed by nitrous air as before. The fame was the cafe with a mixture of acid air and common air.

Having mixed air that had been diminifhed by the fermentation of a mixture of iron filings and brimftone with alkaline air, the water ab-forbed the latter, but left the former, with re-fpect to the teft of nitrous air (and therefore, as I conclude, with refpect to all its properties) the fame that it was before.

Spirit of wine imbibes alkaline air as readily as water, and feems to be as inflammable after-wards as before.

Alkaline air contracts no union with *olive oil*. They were in contact almoft two days, without any diminution of the air. Oil of turpentine, and effential oil of mint, abforbed a very fmall
quantity

2

quantity of alkaline air, but were not fenfibly changed by it.

Ether, however, imbibed alkaline air pretty freely ; but it was afterwards as inflammable as before, and the colour was not changed. It alfo evaporated as before, but I did not attend to this laft circumftance very accurately.

Sulphur, nitre, common falt, and *flints,* were put to alkaline air without imbibing any part of it ; but *charcoal, fpunge,* bits of *linen cloth,* and other fubftances of that nature, feemed to con- denfe this air upon their furfaces ; for it began to diminifh immediately upon their being put to it ; and when they were taken out the alkaline fmell they had contracted was fo pungent as to be almoft intolerable, efpecially that of the fpunge. Perhaps it might be of ufe to recover perfons from fwooning. A bit of fpunge, about as big as a hazel nut, prefently imbibed an ounce meafure of alkaline air.

A piece of the infpiffated juice of *turnfole* was made very dry and warm, and yet it im- bibed a great quantity of the air ; by which it contracted a moft pungent fmell, but the co- lour of it was not changed.

Alum

Alum undergoes a very remarkable change by the action of alkaline air. The outward ſhape and ſize remain the ſame, but the internal ſtructure is quite changed, becoming opaque, beautifully white, and, to appearance, in all reſpects, like alum which had been roaſted ; and ſo as not to be at all affected by a degree of heat that would have reduced it to that ſtate by roaſting. This effect is produced ſlowly ; and if a piece of alum be taken out of alkaline air before the operation is over, the inſide will be tranſparent, and the outſide, to an equal thickneſs, will be a white cruſt.

I imagine that the alkaline vapour ſeizes upon the water that enters into the conſtitution of crude alum, and which would have been expelled by heat. Roaſted alum alſo imbibes alkaline air, and, like the raw alum that has been expoſed to it, acquires a taſte that is peculiarly diſagreeable.

Phoſphorus gave no light in alkaline air, and made no laſting change in its dimenſions. It varied, indeed, a little, being ſometimes increaſed and ſometimes diminiſhed, but after a day and a night, it was in the ſame ſtate as at the firſt. Water abſorbed this air juſt as if nothing had been put to it.

Having

Having put fome *fpirit of falt* to alkaline air, the air was prefently abforbed, and a little of the white falt abovementioned was formed. A little remained unabforbed, and tranfparent, but upon the admiffion of common air to it, it inftantly became white.

Oil of vitriol, alfo formed a white falt with alkaline air, and this did not rife in white fumes.

Acid air, as I have obferved in my former papers, extinguifhes a candle. Alkaline air, on the contrary, I was furprized to find, is flightly inflammable ; which, however, feems to con-firm the opinion of chemifts, that the volatile alkali contains phlogifton.

I dipped a lighted candle into a tall cylindri-cal veffel, filled with alkaline air, when it went out three or four times fucceffively ; but at each time the flame was confiderably enlarged, by the addition of another flame, of a pale yellow colour ; and at the laft time this light flame de-fcended from the top of the veffel to the bot-tom. At another time, upon prefenting a lighted candle to the mouth of the fame veffel, filled with the fame kind of air, the yellowifh flame afcended two inches higher than the flame of the candle. The electric fpark taken in alkaline

kaline air is red, as it is in common inflammable air.

Though alkaline air be inflammable, it appeared, by the following experiment, to be
heavier than the common inflammable air, as
well as to contract no union with it. Into a
veffel containing a quantity of inflammable air,
I put half as much alkaline air, and then about
the fame quantity of acid air. Thefe immediately formed a white cloud, but it did not rife
within the fpace that was occupied by the inflammable air; fo that this latter had kept its
place above the alkaline air, and had not mixed
with-it.

That alkaline air is lighter than acid air is
evident from the appearances that attend the
mixture, which are indeed very beautiful. When
acid air is introduced into a veffel containing
alkaline air, the white cloud which they form
appears at the bottom only, and afcends gradually. But when the alkaline air is put to the
acid, the whole becomes immediately cloudy,
quite to the top of the veffel.

In the laft place, I fhall obferve that alkaline
air, as well as acid, diffolves *ice* as faft as a hot
fire can do it. This was tried when both the
kinds of air, and every inftrument made ufe of
in

in the experiment, had been expofed to a pretty intenfe froft feveral hours. In both cafes, alfo, the water into which the ice was melted diffolved more ice, to a confiderable quantity.

S E C T I O N II.

Of COMMON AIR *diminifhed and made noxious by various proceffes.*

It will have been obferved that, in the firft publication of my papers, I confined myfelf chiefly to the narration of the new *faEts* which I had difcovered, barely mentioning any *hypothefes* that occurred to me, and never feeming to lay much ftrefs upon them. The reafon why I was fo much upon my guard in this refpeEt was, left, in confequence of attaching myfelf to any hypothefis too foon, the fuccefs of my future inquiries might be obftruEted. But fubfequent experiments having thrown great light upon the preceding ones and having confirmed the few conjeEtures I then advanced, I may now venture to fpeak of my hypothefes with a little lefs diffidence. Still, however, I fhall be ready to relinquifh any notions I may now entertain, if new faEts fhould hereafter appear not to favour them.

In

In a great variety of cafes I have obferved
that there is a remarkable *diminution* of common, or refpirable air, in proportion to which
it is always rendered unfit for refpiration, in-
difpofed to effervefce with nitrous air, and incapable of farther diminution from any other
caufe. The circumftances which produce this
effect I had then obferved to be the burning of
candles, the refpiration of animals, the putrefaction of vegetables or animal fubftances,
the effervefcence of iron nlings and brimftone,
the calcination of metals, the fumes of charcoal, the effluvia of paint made of white-lead
and oil, and a mixture of nitrous air.

All thefe proceffes, I obferved, agree in this
one circumftance, and I believe in no other,
that the principle which the chemifts call *phlo-
giflon* is fet loofe ; and therefore I concluded
that the diminution of the air was, in fome
way or other, the confequence of the air becoming overcharged with phlogifton,* and that
water, and growing vegetables, tend to reftore
                                          this

On this account, if it was thought convenient to introduce a new term (or rather make a new application of
a term already in ufe among chemifts) it might not be a-
mifs to call air that has been diminifhed, and made noxious
by any of the proceffes above mentioned, or others fimilar
to them by   common appellation of *phlogifticated air* ;
                                          and,

this air to a ſtate fit for reſpiration, by imbibing the ſuperfluous phlogiſton. Several experiments which I have ſince made tend to con-
firm this ſuppoſition.

Common air, I find, is diminiſhed, and rendered noxious, by *liver of ſelphur*, which the
chemiſts ſay exhales phlogiſton, and nothing
elſe. The diminution in this caſe was one
fifth of the whole, and afterwards, as in other
ſimilar caſes, it made no efferveſcence with nitrous air.

I found alſo, after Dr. Hales, that air is diminiſhed by *Homberg's pyrophorus*.

The ſame effect is produced by firing *gunpowder* in air. This I tried by firing the gunpowder in a receiver half exhauſted, by which
the air was rather more injured than it would
have been by candles burning in it.

Air is diminiſhed by a cement made with one
half common coarſe turpentine and half bees-
wax. This was the reſult of a very caſual ob-

and, if it was neceſſary, the particular proceſs by which it
was phlogiſticated might be added ; as common air phlogiſticated by charcoal, air phlogiſticated by the calcination
of metals, nitrous air phlogiſticated with the liver of ſulphur, &c.

ſervation.

fervation. Having, in an air-pump of Mr.
Smeaton's conftruction, clofed that end of the
fyphon-gage, which is expofed to the outward
air, with this cement (which I knew would
make it perfectly air-light) inftead of fealing it
hermetically ; I obferved that, in a courfe of
time, the quickfilver in that leg kept continually rifing, fo that the meafures I marked
upon it were of no ufe to me; and when I
opened that end of the tube, and clofed it
again, the fame confequence always took place.
At length, fufpecting that this effect muft have
arifen from the bit of *cement* diminifhing the air
to which it was expofed, I covered all the infide
of a glafs tube with it, and one end of it being quite clofed with the cement, I fet it per-
pendicular, with its open end immerfed in a
bafon of quickfilver ; and was prefently fatisfied that my conjecture was well founded : for,
in a few days, the quickfilver rofe fo much
within the tube, that the air in the infide appeared to be diminifhed about one fixth.

To change this air I filled the tube with
quickfilver, and pouring it out again, I re-
placed the tube in its former fituation ; when
the air was diminifhed again, but not fo faft as
before. The fame lining of cement diminifhed the air a third time. How long it will retain this power I cannot tell. This cement had
been

been made feveral months before I made this
experiment with it. I muſt obſerve, however,
that another quantity of this kind of cement,
made with a finer and more liquid turpentine,
had not the power of diminiſhing air, except
in a very fmall proportion. Alſo the common
red cement has this property in the fame fmall
degree. Common air, however, which had been
confined in a glaſs veſſel lined with this cement
about a month, was ſo far injured that a candle
would not burn in it. In a longer time it
would, I doubt not, have become thoroughly
noxious.

Iron that has been ſuffered to ruſt in nitrous
air diminiſhes common air very faſt, as I ſhall
have occaſion to mention when I give a con-
tinuation of my experiments on nitrous air.

Laſtly, the fame effect, I find, is produced
by the *electric ſpark*, though I had no expectation of this event when I made the experiment.

This experiment, however, and thoſe which
I have made in purſuance of it, has fully confirmed another of my conjectures, which relates to the *manner* in which air is diminiſhed
by being overcharged with phlogiſton, viz. the
phlogiſton having a nearer affinity with ſome of
the conſtituent parts of the air than the fixed

N 3   air

air which enters into the compofition of it, in confequence of wnich the fixed air is precipitated.

This I firft imagined from perceiving that lime-water became turbid by burning candles over it, p. 44. This was alfo the cafe with lime-water confined in air in which an animal fubftance was putrefying, or in which an animal died, p. 79. and that in which charcoal was burned, p. 81. But, in all thefe cafes, there was a poffibility of the fixed air being difcharged from the candle, the putrefying fubftance, the lungs of the animal, or the charcoal. That there is a precipitation of lime when nitrous air is mixed with common air, I had not then obferved, but I have fince found it to be the cafe.

That there was no precipitation of lime when brimftone was burned, I obferved, p. 45. might be owing to the fixed air and the lime uniting with the vitriolic acid, and making a falt, which was foluble in water; which falt I, indeed, difcovered by the evaporation of the water.

I alfo obferved, p. 46, 105. that diminifhed air being rather lighter than common air is a circumftance in favour of the fixed, or the
heavier

heavier part of the common air, having been precipitated.

It was upon this idea, together with others fimilar to it, that I took fo much pains to mix fixed air with air diminifhed by refpiration or putrefaction, in order to make it fit for refpiration again ; and I thought that I had, in general, fucceeded to a confiderable degree, p. 99, &c. I will add, alfo, what I did not mention before, that I once endeavoured, but without effect, to preferve mice alive in the fame unchanged air, by fupplying them with fixed air, when the air in which they were confined began to be injured by their refpiration. Without effect, alfo, I confined for fome months, a quantity of quick lime in a given quantity of common air, thinking it might extract the fixed air from it.

The experiments which I made with electricity were folely intended to afcertain what has often been attempted, but, as far as I know, had never been fully accomplifhed, viz. to change the blue colour of liquors, tinged with vegetable juices, red.

For this purpofe I made ufe of a glafs tube, about one tenth of an inch diameter in the infide, as in fig. 16. In one end of this I cemented

N 4   a piece

a piece of wire *b*, on which I put a brafs ball. The lower part from *a* was filled with water tinged blue, or rather purple, with the juice of turnfole, or archil. This is eafily done by an air-pump, the tube being fet in a veffel of the tinged water.

Things being thus prepared, I perceived that, after I had taken the electric fpark, between the wire *b*, and the liquor at *a*, about a minute, the upper part of it began to look red, and in about two minutes it was very manifeftly fo ; and the red part, which was about a quarter of an inch in length, did not readily mix with the reft of the liquor. I obferved alfo, that if the tube lay inclined while I took the fparks, the rednefs extended twice as far on the lower fide as on the upper.

The moft important, though the leaft expected obfervation, however, was that, in proportion as the liquor became red, it advanced nearer to the wire, fo that the fpace of air in which the fparks were taken was diminifhed ; and at length I found that the diminution was about one fifth of the whole fpace ; after which more electrifying produced no fenfible effect.

To determine whether the caufe of the change of colour was in the *air*, or in the *electric matter*,

matter, I expanoed the air which had been diminifhed in the tube by means of an air-pump,
till it expelled all the liquor, and admitted frefh
blue liquor into its place ; but after that, electricity produced no fenfible effect, either on the
air, or on the liquor ; fo that it was evident
that the electric matter had decompofed the air,
and had made it depofite fomething that was of
an acid nature.

In order to determine whether the *wire* had
contributed any thing to this effect, I ufed wires
of different metals, iron, copper, brafs, and
filver ; but the refult was the very fame with
them all.

It was alfo the fame when, by means of a
bent glafs tube, I made the electric fpark without any wire at all, in the following manner.
Each leg of the tube, fig. 19. ftood in a bafon
of quickfilver ; which, by means of an air-
pump, was made to afcend as high as *a, a,* in
each leg, while the fpace between *a* and *b* in
each contained the blue liquor, and the fpace
between *b* and *b* contained common air. Things
being thus dispofed, I made the electric fpark
perform the circuit from one leg to the other,
paffing from the liquor in one leg of the tube
to the liquor in the other leg, through the fpace
of air. The effect was, that the liquor, in
both

both the legs, became red, and the fpace of air between them was contracted, as before.

Air thus diminifhed by electricity makes no effervefcence with, and is no farther diminifhed by a mixture of nitrous air; fo that it muft have been in the higheft degree noxious, exactly like air diminifhed by any other procefs.

In order to determine what the *acid* was, which was depofited by the air, and which changed the colour of the blue liquor, I expofed a fmall quantity of the liquor fo changed to the common air and found that it recovered its blue colour, exactly as water, tinged with the fame blue, and impregnated with fixed air, will do. But the following experiment was ftill more decifive to this purpofe. Taking the electric fpark upon *lime-water*, inftead of the blue liquor, the lime was precipitated as the air diminifhed.

From thefe experiments it pretty clearly follows, that the electric matter either is, or contains phlogifton; fince it does the very fame thing that phlogifton does. It is alfo probable, from thefe experiments, that the fulphureous fmell, which is occafioned by electricity, being very different from that of fixed air, the phlogifton

gifton in the electric matter itself may contribute
to it.

It was now evident that common air diminifhed by any one of the proceffes abovementioned being the fame thing, as I have obferved,
with air diminifhed by any other of them (fince
it is not liable to be farther diminifhed by any
other) the lofs which it fuftains, in all the cafes,
is, in part, that of the *fixed air* which entered
into its conftitution. The fixed air thus precipitated from common air by means of phlogifton unites with lime, if any lime water be
ready to receive it, unlefs there be fome other
fubftance at hand, with which it has a greater
affinity, as the *calces of metals.*

If the whole of the diminution of common
air was produced by the depofition of fixed air,
it would be eafy to afcertain the quantity of
fixed air that is contained in any given quantity
of common air. But it is evident that the
whole of the diminution of common air by
phlogifton is not owing to the precipitation of
fixed air, becaufe a mixture of nitrous air will
make a great diminution in all kinds of air
that are fit for refpiration, even though they
never were common air, and though nothing
was ufed in the procefs for generating them that
can be fuppofed to yield fixed air.

Indeed,

Indeed, it appears, from fome of the experiments, that the diminution of fome of thefe kinds of air by nitrous air is fo great, and approaches fo nearly to the quantity of the diminution of common air by the fame procefs, as to fhew that, unlefs they be very differently affected by phlogifton, very little is to be allowed to the lofs of fixed air in the diminution of common air by nitrous air.

The kinds of air on which this experiment was made were inflammable air, nitrous air diminifhed by iron filings and brimftone, and nitrous air itfelf; all of which are produced by the folution of metals in acids; and alfo on common air diminifhed and made noxious, and therefore deprived of its fixed air by phlogiftic proceffes ; and they were reftored to a great degree of purity by agitation in water, out of which its own air had been carefully boiled.

To five parts of inflammable air, which had been agitated in water till it was diminifhed about one half (at which time part of it fired with a weak explofion) I put one part of nitrous air, which diminifhed it one eighth of the whole. This was done in lime-water, without any precipitation of lime.   To compare this with common air, I mixed the fame quantity, *viz* five parts of this, and one part of nitrous air : when
a con-

confiderable cruft of lime was formed upon
the furface of the lime-water, though the diminution was very little more than in the former
procefs. It is poffible, however, that the common air might have taken more nitrous air
before it was fully faturated, fo as to begin to
receive an addition to its bulk.

I agitated in water a quantity of nitrous air
phlogifticated with iron filings and brimftone,
and found it to be fo far reftored, that three
fourths of an ounce meafure of nitrous air being
put to two ounce meafures of it, made no addition to it.

But the moft remarkable of thefe experiments
is that which I made with *nitrous air* itfelf which
I had no idea of the poffibility of reducing to
a ftate fit for refpiration by any procefs whatever, at the time of my former publication on
this fubject. This air, however, itfelf, without
any previous phlogiftication, is purified by agitation in water till it is diminifhed by frefh
nitrous air, and to a very confiderable degree.

In a pretty long time I agitated nitrous air in
water, fupplying it from time to time with
more, as the former quantity diminifhed, till
only one eighteenth of the whole quantity
remained ; in which ftate it was fo wholefome,
that

that a moufe lived in two ounce meafures of it
more than ten minutes, without fhewing any
fign of uneafinefs ; fo that I concluded it muft
have been about as good as air in which
candles had burned out. After agitating it
again in water, I put one part of frefh nitrous
air to five parts of this air, and it was diminifhed one ninth part. I then agitated it a third
time, and putting more nitrous air to it, it was
diminifhed again in the fame proportion, and fo
a fourth time ; fo that, by continually repeating
the procefs, it would, I doubt not, have been
all abforbed. Thefe proceffes were made in
lime-water, without forming any incruftation
on the furface of it.

Laftly, I took a quantity of common air,
which had been diminifhed and made noxious
by phlogiftic proceffes ; and when it had been
agitated in water, I found that it was diminifhed by nitrous air, though not fo much as it
would have been at the firft. After cleanfing
it a fecond time, it was diminifhed again by the
fame means ; and, after that, a third time ; and
thus there can be no doubt but that, in time,
the whole quantity would have difappeared.
For I have never found that agitation in water,
deprived of its own air, made any addition to
a quantity of noxious air ; though, *a priori*, it
might have been imagined that, as a faturation
with

with phlogifton diminifhes air, the extraction of
phlogifton would increafe the bulk of it. On
the contrary, agitation in water always diminifhed noxious air a little; indeed, if water be
deprived of all its own air, it is impoffible to
agitate any kind of air in it without fome lofs.
Alfo, when noxious air has been reftored by'
plants, I never perceived that it gained any
addition to its bulk by that means. There was
no incruftation of the lime-water in the above-
mentioned experiment.

It is not a little remarkable, that thofe kinds
of air which never had been common air, as
inflammable air, phlogifticated nitrous air, and
nitrous air itfelf, when rendered wholefome by
agitation in water, fhould be more diminifhed
by frefh nitrous air, than common air which
had been made noxious, and reftored by the
fame procefs; and yet, from the few trials that
I have made, I could not help concluding that
this is the cafe.

In this courfe of experiments I was very near
deceiving myfelf, in confequence of transferring
the nitrous air which I made ufe of in a bladder, in the manner defcribed, p. 15. fig. 9. fo
as to conclude that there was a precipitation of
lime in all the above-mentioned cafes, and that
even nitrous air itfelf produced that effect. But
after

after repeated trials, I found that there was no
precipitation of lime, except in the firft diminution of common air, when the nitrous air
was transferred in a glafs veffel.

That the calces of metals contain air, of fome
kind or other, and that this air contributes to
the additional weight of the calces, above that.
of the metals from which they are made, had
been obferved by Dr. Hales ; and Mr. Hartley
had informed me, that when red-lead is boiled
in linfeed oil, there is a prodigious difcharge of
air before they .incorporate.   I had likewife
found, that no weight is either gained or loft
by the calcination of tin in a clofe glafs veffel ;
but I purpofely deferred making any more
experiments on the fubject, till we fhould have
fome weather in which I could make ufe of a
large burning lens, which I had provided for
that and other purpofes ; but, in the mean
time, I was led to the difcovery in a different
manner.

Having, by the laft-re cited experiments, been
led to confider the electric matter as phlogifton,
or fomething containing phlogifton, I was endeavouring to revivify the calx of lead with it ;
when I was furprized to perceive a confiderable
generation of air.   It occurred to me, that poffibly this effect might arife from the *heat* communicated

municated to the red-lead by the electric sparks, and therefore I immediately filled a small phial with the red-lead, and heating it with a candle, I presently expelled from it a quantity of air about four or five times the bulk of the lead, the air being received in a vessel of quicksilver. How much more air it would have yielded, I did not try.

Along with the air, a small quantity of *water* was likewise thrown out ; and it immediately occurred to me, that this water and air together must certainly be the cause of the addition of weight in the calx It still remained to examine what kind of air this was ; but admitting water to it, I found that it was imbibed by it, exactly like *fixed air*, which I therefore immediately concluded it must be *.

After this, I found that Mr. Lavoisier had completely discovered the same thing, though his apparatus being more complex, and less

---

* Here it becomes me to ask pardon of that excellent philosopher Father Beccaria of Turin, for conjecturing that the phlogiston, with which he revivified metals, did not come from the electric matter itself, but from what was discharged from other pieces of metal with which he made the experiment. See History of Electricity, p. 277, &c. This *revivification of metals* by electricity completes the proof of the electric matter being, or containing phlogiston.

O　　　　　accurate

accurate than mine, he concluded that more of
the air difcharged from the calces of metals
was immifcible with water than I found it to
be. It appeared to me that I had never obtained fixed air more pure.

It being now pretty clearly determined, that
common air is made to depofit the fixed air
which entered into the conftitution of it, by
means of phlogifton, in all the cafes of diminiſhed air, it will follow, that in the precipitation of lime, by breathing into lime-water
the fixed air, which incorporates with lime,
comes not from the lungs, but from the common air, decompofed by the phlogifton exhaled
from them, and difcharged, after having been
taken in with the aliment, and having performed its function in the animal fyftem.

Thus my conjecture is more confirmed, that
the caufe of the death of animals in confined
air is not owing to the want of any *pabulum
vitæ*, which the air had been fuppofed to contain, but to the want of a difcharge of the
phlogiftic matter, with which the fyftem was
loaded; the air, when once faturated with it,
being no fufficient *menftruum* to take it up.

The inftantaneous death of animals put into
air fo vitiated, I ftill think is owing to fome
ftimulus,

ſtimulus, which, by cauſing immediate, univer-
ſal and violent convulſions, exhauſts the whole
of the *vis vitæ* at once; becauſe, as I have obſerved, the manner of their death is the very
ſame in all the different kinds of noxious air.

To this ſection on the ſubject of diminiſhed, and noxious air, or as it might have been
called *phlogiſticated air*, I ſhall ſubjoin a letter
which I addreſſed to Sir John Pringle, on the
noxious quality of the effluvia of putrid marſhes,
and which was read at a meeting of the Royal
Society, December 16, 1773.

This letter which is printed in the Philoſophical Tranſactions, Vol. 74, p. 90. is immediately followed by another paper, to which I would
refer my reader. It was written by Dr. Price,
who has ſo greatly diſtinguiſhed himſelf, and
done ſuch eminent ſervice to his country, and
to mankind, by his calculations relating to the
probabilities of human life, and was ſuggeſted
by his hearing this letter read at the Royal Society. It contains a confirmation of my obſervations on the noxious effects of ſtagnant waters
by deductions from Mr. Muret's account of the
Bills of Mortality for a pariſh ſituated among
marſhes, in the diſtrict of Vaud, belonging to
the Canton of Bern in Switzerland.

To

To Sir J O H N P R I N G L E, Baronet.

D E A R S I R,

Having purſued my experiments on different
kinds of air conſiderably farther, in ſeveral reſpects, than I had done, when I preſented the
laſt account of them to the Royal Society;
and being encouraged by the favourable notice
which the Society has been pleaſed to take of
them, I ſhall continue my communications on
this ſubject; but, without waiting for the reſult of a variety of proceſſes, which I have
now going on, or of other experiments, which
I propoſe to make, I ſhall, from time to time,
communicate ſuch detached articles, as I. ſhall
have given the moſt attention to, and with reſpect to which, I ſhall have been the moſt ſucceſsful in my inquiries.

Since the publication of my papers, I have
read two treatiſes, written by Dr. Alexander,
of Edinburgh, and am exceedingly pleaſed
with the ſpirit of philoſophical inquiry, which
they diſcover. They appear to me to contain
many new, curious, and valuable obſervations;
but one of the *concluſions*, which he draws
from his experiments, I am ſatisfied, from my
own

own obfervations, is ill founded, and from the
nature of it, muft be dangerous. I mean his
maintaining, that there is nothing to be apprehended from the neighbourhood of putrid
marfhes.

I was particularly furprifed, to meet with
fuch an opinion as this, in a book infcribed to
yourfelf, who have fo clearly explained the
great mifchief of fuch a fituation, in your excellent treatife *on the difeafes of the army.* On
this account, I have thought it not improper,
to addrefs to you the following obfervations
and experiments, which I think clearly demonftrate the fallacy of Dr. Alexander's reafon-
ing, indifputably eftablifh your doctrine, and
indeed juftify the apprehenfions of all mankind
in this cafe.

I think it probable enough, that putrid matter, as Dr. Alexander has endeavoured to
prove, will preferve other fubftances from putrefaction ; becaufe, being already faturated
with the putrid effluvium, it cannot readily
take any more ; but Dr. Alexander was not
aware, that air thus loaded with putrid effluvium is exceedingly noxious when taken into
the lungs. I have lately, however, had an opportunity of fully afcertaining how very noxious
fuch air is.

Hap-

Happening to ufe at Calne, a much larger trough of water, for the purpofe of my experiments, than I had done at Leeds, and not having frefh water fo near at hand as I had there, I neglected to change it, till it turned black, and became offenfive, but by no means to fuch a degree, as to deter me from making ufe of it. In this ftate of the water, I obferved bubbles of air to rife from it, and efpecially in one place, to which fome fhelves, that I had in it, directed them ; and having fet an inverted glafs veffel to catch them, in a few days I collected a confiderable quantity of this air, which iffued fpontaneoufly from the putrid water; and putting nitrous air to it, I found that no change of colour or diminution enfued, fo that it muft have been, in the higheft degree, noxious. I repeated the fame experiment feveral times afterwards, and always with the fame refult.

After this, I had the curiofity to try how wholefome air would be affected by this water ; when, to my real furprife, I found, that after only one minute's agitation in it, a candle would not burn in it ; and, after three or four minutes, it was in the fame ftate with the air, which had iffued fpontaneoufly from the fame water,

I alfo

I alfo found, that common air, confined in a glafs veffel, in *contact* only with this water, and without any agitation, would not admit a candle to burn in it after two days.

Thefe facts certainly demonftrate, that air which either arifes from ftagnant and putrid water, or which has been for fome time in contact with it, muft be very unfit for refpiration ; and yet Dr. Alexander's opinion is rendered fo plaufible by his experiments, that it is very poffible that many perfons may be rendered fecure, and thoughtlefs of danger, in a fituation in which they muft neceffarily breathe it. On this account, I have thought it right to make this communication as early as I conveniently could ; and as Dr. Alexander appears to be an ingenious and benevolent man, I doubt not but he will thank me for it.

That air iffuing from water, or rather from the foft earth, or mud, at the bottom of pits containing water, is not always unwholefome, I have alfo had an opportunity of afcertaining. Taking a walk, about two years ago, in the neighbourhood of Wakefield, in Yorkfhire, I obferved bubbles of air to arife, in remarkably great plenty, from a fmall pool of water, which, upon inquiry, I was informed had been the place, where fome perfons had been boring the

O 4 ground,

ground, in order to find coal. Thefe bubbles of air having excited my curiofity, I prefently returned, with a bafoh, and other veffels proper for my purpofe, and having ftirred the mud with a long ftick, I foon got about a pint of this air; and, examining it, found it to be good common air; at leaft a candle burned in it very well. I had not then difcovered the method of afcertaining the goodnefs of common air, by a mixture of nitrous air.ᶜ Previous to the trial, I had fufpeéted that this air would have been found to be inflammable.

I fhall conclude this letter with obferving, that I have found a remarkable difference in different kinds of water, with refpeét to their effeét on common air agitated in them, and which I am not yet able to account for. If I agitate common air in the water of a deep well, near my houfe in Calne, which is hard, but clear and fweet, a candle will not burn in it after three minutes. The fame is the cafe with the rain-water, which I get from the roof of my houfe. But in diftilled water, or the water of a fpring-well near the houfe, I muft agitate the air about twenty minutes, before it will be fo much injured. It may be worth while, to make farther experiments with refpeét to this property of water.

In

In confequence of ufing the rain-water, and the well-water above mentioned, I was very near concluding, contrary to what I have af-ferted in this treatife, that common air fuf-fers a decompofition by great rarefaction. For when I had collected a confiderable quantity of air, which had been rarefied about four hundred times, by an excellent pump made for me by Mr. Smeaton, I always found, that if I filled my receivers with the water above mentioned, though I did it fo gradually as to occafion as little agitation as poffible, a candle would not burn in the air that remained in them. But when I ufed diftilled water, or frefh fpring-water, I undeceived myfelf.

I think myfelf honoured by the attention, which, from the firft, you have given to my experiments, and am, with the greateft refpect,

Dear Sir,

Your moft obliged

Humble Servant,

London, 7 Dec. 1773.

J. PRIESTLEY.

POST-

POSTSCRIPT.

I cannot help expreffing my furprize, that
fo clear and intelligible an account, of Mr.
SMEATON's air-pump, fhould have been before
the public fo long, as ever fince the publication
of the forty-feventh volume of the Philofophical Tranfactions, printed in 1752, and yet that
none of our philofophical inftrument-makers
fhould ufe the conftruction. The fuperiority
of this pump, to any that are made upon the
common plan, is, indeed, prodigious. Few of
them will rarefy more than 100 times, and, in
a general way, not more than 60 or 70 times;
whereas this inftrument muft be in a poor ftate
indeed, if it does not rarefy 200 or 300 times;
and when it is in good order, it will go as far as
1000 times, and fometimes even much farther
than that; befides, this inftrument is worked
with much more eafe, than a common air-
pump, and either exhaufts or condenfes at pleafure. In fhort, to a perfon engaged in philofophical purfuits, this inftrument is an invalu-
able acquifition. I fhall have occafion to recite
fome experiments, which I could not have made,
and which, indeed, I fhould hardly have dared
to attempt, if I had not been poffeffed of fuch
an air-pump as this. It is much to be wifhed,
that fome erfon of fpirit in the trade would at-
                                                    tempt

tempt the conftruction of an inftrument, which would do great credit to himfelf, as well as be of eminent fervice to philofophy.

SECTION III.

Of Nitrous Air.

Since the publication of my former papers I have given more attention to the fubject of nitrous air than to any other fpecies of air ; and having been pretty fortunate in my inquiries, I fhall be able to lay before my reader a more fatisfactory account of the curious phenomena occafioned by it, and alfo of its nature and conftitution, than I could do before, though much ftill remains to be inveftigated concerning it, and many new objects of inquiry are ftarted.

With a view to difcover where the power of nitrous air to diminifh common air lay, I evaporated to drynefs a quantity of the folution of copper in diluted fpirit of nitre ; and having procured from it a quantity of a *green precipitate*, I threw the focus of a burning-glafs upon it, when it was put into a veffel of quickfilver, ftanding inverted in a bafon of quickfilver. In this manner I procured air from it, which

which appeared to be, in all reſpects, nitrous
air; ſo that part of the ſame principle which
had eſcaped during the ſolution, in the form of
air, had likewiſe been retained in it, and had
not left it in the evaporation of the water.

With great difficulty I alſo procured a ſmall
quantity of the ſame kind of air from a ſolution of *iron* in ſpirit of nitre, by the ſame proceſs.

Having, for a different purpoſe, fired ſome
paper, which had been dipped in a ſolution of
copper in diluted ſpirit of nitre, in nitrous air,
I found there was a conſiderable addition to the
quantity of it; upon which I fired ſome of the
ſame kind of paper in quickſilver and preſently
obſerved that air was produced from it in great
plenty. This air, at the firſt, ſeemed to have
ſome ſingular properties, but afterwards I found
that it was nothing more than a mixture of nitrous air, from the precipitate of the ſolution,
and of inflammable air, from the paper; but
that the former was predominant.

In the mixture of this kind of air with common air, in a trough of water which had been
putrid, but which at that time ſeemed to have
recovered its former ſweetneſs (for it was not in
the leaſt degree offenſive to the ſmell) a phenomenon

menon fometimes occurred, which for a long
time exceedingly delighted and puzzled me ;
but which was afterwards the means of letting
me fee much farther into the conftitution of nitrous air than I had been able to fee before.

When the diminution of the air was nearly
completed, the veffel in which the mixture was
made began to be filled with the moft beautiful
white fumes, exactly refembling the precipitation
of fome white fubftance in a tranfparent menftruum, or the falling of very fine fnow ; except that it was much thicker below than above,
as indeed is the cafe in all chemical precipita-
tions. This appearance continued two or three
minutes.

At other times I went over the fame procefs,
as nearly as poffible in the fame manner, but
without getting this remarkable appearance, and
was feveral times greatly difappointed and cha-
grined, when I baulked the expectations of my
friends, to whom I had defcribed, and meant to
have fhewn it. This made me give all the attention I poffibly could to this experiment, endeavouring to recollect every circumftance,
which, though unfufpected at the time, might
have contributed to produce this new appearance ; and I took a great deal of pains to procure a quantity of this air from the paper
above

above mentioned for the purpofe, which, with
a fmall burning lens, and an uncertain fun, is
not a little troublefome. But all that I obferved
for fome time was, that I ftood the beft chance
of fucceeding when I *warmed* the veffel in which
the mixture was made, and *agitated* the air during the effervefcence.

Finding, at length, that, with the fame
preparation and attentions, I got the fame appearance from a mixture of nitrous and common air in the fame trough of water, I concluded that it could not depend upon any
thing peculiar to the precipitate of the *copper*
contained in the *paper* from which the air was
procured, as I had at firft imagined, but upon
what was common to it, and pure nitrous
air.

Afterwards, having, (with a view to obferve
whether any cryftals would be formed by the
union of volatile alkali, and nitrous air, fimilar
to thofe formed by it and fixed air, as defcribed
by Mr. Smeth in his *Differtation on fixed Air*)
opened the mouth of a phial which was half
filled with a volatile alkaline liquor, in a jar of
nitrous air (in the manner defcribed p. 11.
fig. 4.) I had an appearance which perfectly
explained the preceding. All that part of the
phial which was above the liquor, and which
contained

contained common air, was filled with beautiful *white clouds*, as if fome fine white powder had
been inftantly thrown into it, and fome of thefe
clouds rofe within the jar of nitrous air. This
appearance continued about a minute, and then
intirely difappeared, the air becoming tranfparent.

Withdrawing the phial, and expofing it to
the common air, it there alfo became turbid,
and foon after the tranfparency returned. Introducing it again into the nitrous air, the
clouds appeared as before. In this manner the
white fumes, and tranfparency, fucceeded each
other alternately, as often as I chofe to repeat
the experiment, and would no doubt have continued till the air in the jar had been thoroughly
diluted with common air. Thefe appearances
were the fame with any fubftance that contained
volatile alkali, fluid or folid.

When, inftead of the fmall phial, I ufed a
large and tall glafs jar, this appearance was
truly fine and ftriking, efpecially when the water in the trough was very tranfparent. For I
had only to put the fmalleft drop of a volatile
alkaline liquor, or the fmalleft bit of the folid
falt, into the jar, and the moment that the
mouth of it was opened in a jar of nitrous
air, the white clouds above mentioned began to
be

be formed at the mouth, and prefently de-
fcended to the bottom, fo as to fill the whole,
were it ever fo large, as with fine fnow.

In confidering this experiment, I foon perceived that this curious appearance muft have
been occafioned by the mixture of the nitrous
and common air, and therefore that the white
clouds muft be *nitrous ammoniac*, formed by the
acid of the nitrous air, fet loofe in the decompofition of it by common air, while the phlogifton, which muft be another conftituent part
of nitrous air, entering the common air, is the
caufe of the diminution it fuffers in this procefs ;
as it is the caufe of a fimilar diminution, in a
variety of other procefſes.

I would obferve, that it is not peculiar to nitrous air to be a teft of the fitnefs of air for refpiration. Any other procefs by which air is
diminifhed and made noxious anfwers the fame
purpofe. Liver of fulphur for inftance, the
calcination of metals, or a mixture of iron
filings and brimftone will do juft the fame
thing ; but the application of them is not fo
eafy, or elegant, and the effect is not fo foon
perceived. In fact, it is *phlogiſton* that is the
teft. If the air be fo loaded with this principle
that it can take no more, which is feen by its
not being diminifhed in any of the procefſes
                                        above

above mentioned, it is noxious ; and it is wholefome in próportion to the quantity of phlogifton
that it is able to take.

This, I have no doubt, is the true theory of
the diminution of common air by nitrous air,
the rednefs of the appearance being nothing
more than the ufual colour of the fumes of
fpirit of nitre, which is now difengaged from
the fuperabundant phlogifton with which it
was combined in the nitrous air, and ready to
form another union with any thing that is at
hand, and capable of it.

With the volatile alkali it forms nitrous ammoniac, water imbibes it like any other acid,
even quickfilver is corroded by it ; but this
action being flow, the rednefs in this mixture of
nitrous and common air continues much longer
when the procefs is made in quickfilver, than
when it is made in water, and the diminution, as
I have alfo obferved, is by no means fo great.

I was confirmed in this opinion when I put
a bit of volatile alkaline falt into the jar of
quickfilver in which I made the mixture of
nitrous and common air. In thefe circumftances, the veffel being previoufly filled with
the alkaline fumes, the acid immediately joined
them, formed the white clouds above mention-
P   ed,

ed, and the diminution proceeded almoft as far as when the procefs was made in water. That it did not proceed quite fo far, I attribute chiefly to the fmall quantity of calx formed by the flight folution of mercury with the acid fumes not being able to abforb all the fixed air that is precipitated from the common air by the phlogifton.

In part, alfo, it may be owing to the fmall quantity of furface in the quickfilver in the veffels that I made ufe of ; in confequence of which the acid fumes could act upon it only in a flow fucceffion, fo that part of them, as well as of the fixed air, had an opportunity of forming another union with the dimi-nifhed air.

This, as I have obferved before, was fo much the cafe when the procefs was made in quick-filver, without any volatile alkali, that when water was admitted to it, after fome time, it was not capable of diffolving that union, tho' it would not have taken place if the procefs had been in water from the firft.

In diverfifying this experiment, I found that it appeared to very great advantage when I fufpended a piece of volatile falt in the com-mon air, previous to the admiffion of nitrous
air

air to it, inclofing it in a bit of gauze, muflin, or a fmall net of wire. For, prefently after the rednefs of the mixture begins to go off, the white cloud, like fnow, begins to defcend from the falt, as if a white powder was fhaken out of the bag that contains it. This white cloud prefently fills the whole veffel, and the appearance will laft about five minutes.

If the falt be not put to the mixture of thefe two kinds of air till it has perfectly recovered its tranfparency, the effervefcence being completely over, no white cloud will be formed; and, what is rather more remarkable, there is nothing of this appearance when the falt is put into the nitrous air itfelf. The reafon of this muft be, that the acid of the nitrous air has a nearer affinity with its phlogifton than with the volatile alkali; though, the phlogifton having a nearer affinity with fomething in the common air, the acid being thereby fet loofe, will unite with the alkaline vapour, if it be at hand to unite with it.

There is alfo very little of any white cloud formed upon holding a piece of the volatile falt within the mouth of a phial containing fmoking fpirit of nitre. Alfo when I threw the focus of a burning mirror upon fome fal ammoniac in nitrous air, and filled the whole

veffel

veffel with white fumes which arofe from it, they were foon difperfed, and the air was neither diminifhed nor altered.

I was now fully convinced, that the white cloud which I cafually obferved, in the firft of thefe experiments, was occafioned by the volatile alkali emitted from the water, which was in a flight degree putrid; and that the warming, and agitation of the veffels, had promoted the emiffion of the putrid, or alkaline effluvium.

I could not perceive that the diminution of common air by the mixture of nitrous air was fenfibly increafed by the prefence of the volatile alkali. It is poffible, however, that, by affifting the water to take up the acid, fomething lefs of it may be incorporated with the remaining diminifhed air than would otherwife have been; but I did not give much attention to this circumftance.

When the phial in which I put the alkaline falts contained any kind of noxious air, the opening of it in nitrous air was not followed by any thing of the appearance above mentioned. This was the cafe with inflammable air. But when, after agitating the inflammable air in water, I had brought it to a ftate in which it

was

was diminifhed a little by the mixture of nitrous air, the cloudy appearance was in the fame proportion; fo that this appearance feems to be equally a teft of the fitnefs of air for refpiration, with the rednefs which attends the mixture of it with nitrous air only.

Having generally faftened the fmall bag which contained the volatile falt to a piece of brafs wire in the preceding experiment, I commonly found the end of it corroded, and covered with a blue fubftance. Alfo the falt itfelf, and fometimes the bag was died blue. But finding that this was not the cafe when I ufed an iron wire in the fame circumftances, but that it became *red*, I was fatisfied that both the metals had been diffolved by the volatile alkali. At firft I had a fufpicion that the blue might have come from the copper, out of which the nitrous air had been made. But when the nitrous air was made from iron, the appearances were, in all refpects, the fame.

I have obferved, in the preceding fection, that if nitrous air be mixed with common air in *lime-water*, the furface of the water, where it is contiguous to that mixture, will be covered with an incruftation of lime, fhewing that fome fixed air had been depofited in the procefs. It is remarkable, however, as I there alfo juft

P 3   mentioned,

mentioned, that this is the cafe when nitrous
air alone is put to a veffel of lime-water, after
it has been kept in a *bladder*, or only transferred
from one veffel to another by a bladder, in the
manner defcribed, p. 15. fig. 9.

As I had ufed the fame bladder for transferring various kinds of air, and among the reft
fixed air, I firft imagined that this effect might
have been occafioned by a mixture of this fixed
air with the nitrous air, and therefore took a
frefh bladder ; but ftill the effect was the fame.
To fatisfy myfelf farther, that the bladder had
produced this effect, I put one into a jar of
nitrous air, and after it had continued there a
day and a night, I found that the nitrous air in
this jar, though it was transferred in a glafs
veffel, made lime-water turbid.

Whether there was any thing in the preparation of thefe bladders that occafioned their producing this effect, I cannot tell. They were
fuch as I procure from the apothecaries. The
thing feems to deferve farther examination, as
there feems, in this cafe, to be the peculiar effect
of fixed air from other caufes, or elfe a production of fixed air from materials that have
not been fuppofed to yield it, at leaft not in
circumftances fimilar to thefe.

As

As fixed air united to water diffolves iron, I had the curiofity to try whether fixed air alone would do it ; and as nitrous air is of an *acid* nature, as well as fixed air, I, at the fame time, expofed a large furface of iron to both the kinds ; firft filling two eight ounce phials with nails, and then with quickfilver, and after that difplacing the quickfilver in one of the phials by fixed air, and in the other by nitrous air ; then inverting them, and leaving them with their mouths immerfed in bafons of quickfilver.

In thefe circumftances the two phials ftood about two months, when no fenfible change at all was produced in the fixed air, or in the iron which had been expofed to it, but a moft remarkable, and moft unexpected change was made in the nitrous air ; and in purfuing the experiment, it was transformed into a fpecies of air, with properties which, at the time of my firft publication on this fubject, I fhould not have hefitated to pronounce impoffible, *viz.* air in which a candle burns quite naturally and freely, and which is yet in the higheft degree noxious to animals, infomuch that they die the moment they are put into it ; whereas, in general, animals live with little fenfible inconvenience in air in which candles have burned out. Such, however, is nitrous air, after it has been long expofed to a large furface of iron.

It

It is not lefs extraordinary, that a ftill longer continuance of nitrous air in thefe circumftances (but *how long* depends upon too many, and too minute circumftances to be afcertained with exactnefs) makes it not only to admit a candle to burn in it, but enables it to burn with an *enlarged flame*, by another flame (extending every where to an equal diftance from that of the candle, and often plainly diftinguifhable from it) adhering to it. Sometimes I have perceived the flame of the candle, in thefe circumftances, to be twice as large as it is naturally, and sometimes not lefs than five or fix times larger; and yet without any thing like an *explofion*, as in the firing of the weakeft inflammable air.

Nor is the farther progrefs in the tranfmutation of nitrous air, in thefe circumftances, lefs remarkable. For when it has been brought to the ftate laft mentioned, the agitation of it in frefh water almoft inftantly takes off that peculiar kind of inflammability, fo that it extinguifhes a candle, retaining its noxious quality. It alfo retains its power of diminifhing common air in a very great degree.

But this noxious quality, like the noxious quality of all other kinds of air that will bear agitation in water, is taken out of it by this operation, continued about five minutes; in
which

which procefs it fuffers a farther and very confiderable diminution. It is then itfelf diminifhed by frefh nitrous air, and animals live in
it very well, about as well as in air in which
candles have burned out.

Laftly, One quantity of nitrous air, which
had been expofed to iron in quickfilver, from
December 18 to January 20, and which happened to ftand in water till January 31 (the iron
ftill continuing in the phial) was fired with an
explofion, exactly like a weak inflammable air.
At the fame time another quantity of nitrous
air, which had likewife been expofed to iron,
ftanding in quickfilver, till about the fame time,
and had then ftood in water only, without iron,
only admitted a candle to burn in it with an
enlarged flame, as in the cafes above mentioned.
But whether the difference I have mentioned in
the circumftances of thefe experiments contributed to this difference in the refult, I cannot
tell.

Nitrous air treated in the manner above mentioned is diminifhed about one fourth by ftanding in quickfilver; and water admitted to it will
abforb about half the remainder; but if water
only, and no quickfilver, be ufed from the beginning, the nitrous air will be diminifhed much
fafter and farther; fo that not more than one
fourth,

fourth, one fixth, or one tenth of the original quantity will remain. But I do not know that there is any difference in the conftitution of the air which remains in thefe two cafes.

The water which has imbibed this nitrous air expofed to iron is remarkably green, alfo the phial containing it becomes deeply, and, I be-lieve, indelibly tinged with green ; and if the water be put into another veffel, it prefently depofits a confiderable quantity of matter, which when dry appears to be the earth or ochre of iron ; from which it is evident, that the acid of the nitrous air diffolves the iron ; while the phlo-gifton, being fet loofe, diminifhes nitrous air, as in the procefs of the iron filings and brim-ftone.

Upon this hint, inftead of ufing *iron*, I in-troduced a pot of *liver of fulphur* into a jar of nitrous air, and prefently found, that what I had before done by means of iron in fix weeks, or two months, I could do by liver of fulphur (in confequence, no doubt, of its giving its phlogifton more freely) in lefs than twenty-four hours, efpecially when the procefs was kept warm.

It is remarkable, however, that if the procefs with liver of fulphur be fuffered to proceed,
the

the nitrous air will be diminifhed much farther. At one time not more than one twentieth of the original quantity remained, and how much farther it might have been diminifhed, I cannot tell. In this great diminution, it does not admit a candle to burn in it at all ; and I generally found this to be the cafe whenever the diminution had proceeded beyond three fourths of the original quantity *.

It is fomething remarkable, that though the diminution of nitrous air by iron filings and brimftone very much refembles the diminution of it by iron only, or by liver of fulphur, yet the iron filings and brimftone never bring it to fuch a ftate as that a candle will burn in it ; and alfo that, after this procefs, it is never capable of diminifhing common air. But when it is confidered that thefe properties are deftroyed by agitation in water, this difference in the refult of procefles, in other refpects fimilar, will appear lefs extraordinary ; and they agree in this, that long agitation in water makes both thefe kinds of nitrous air equally fit for refpiration, being equally diminifhed by frefh nitrous air. It is poflible that there would have been

* The refult of feveral of thefe experiments I had the pleafure of trying in the prefence of the celebrated Mr. De Luc of Geneva, when he was upon a vifit to Lord Shelburne in Wiltfhire.

a more

a more exaćt agreement in the refult of thefe
proceffes, if they had been made in equal degrees of *heat* ; but the procefs with iron was
made in the ufual temperature of the atmofphere, and that with liver of fulphur generally
near a fire.

It may clearly, I think, be inferred from thefe
experiments, that all the difference between
frefh nitrous air, that ftate of it in which it is
partially inflammable, or wholly fo, that in
which it again extinguifhes candles, and that in
which it finally becomes fit for refpiration, depends upon fome difference in the *mode of the*
combination of its acid with phlogifton, or on
the *proportion* between thefe two ingredients in
its compofition ; and it is not improbable but
that, by a little more attention to thefe experiments, the whole myftery of this proportion
and combination may be explained.

I muft not omit to obferve that there was
fomething peculiar in the refult of the firft experiment which I made with nitrous air expofed to iron ; which was that, without any agitation in water, it was diminifhed by frefh nitrous air, and that a candle burned in it quite
naturally. To what this difference was owing
I cannot tell. This air, indeed, had been expofed to the iron a week or two longer than in
any

any of the other cafes, but I do not imagine that this circumftance could have produced that difference.

When the procefs is in water with iron, the time in which the diminution is accomplifhed is exceedingly various ; being fometimes completed in a few days, whereas at other times it has required a week or a fortnight. Some kinds of iron alfo produced this effect much fooner than others, but on what circumftances this difference depends I do not know. What are the varieties in the refult of this experiment when it is made in quickfilver I cannot tell, becaufe, on account of its requiring more time, I have not repeated it fo often ; but I once found that nitrous air was not fenfibly changed by having been expofed to iron in quickfilver nine days ; whereas in water a very confiderable alteration was always made in much lefs than half that time.

It may juft deferve to be mentioned, that nitrous air extremely rarified in an air-pump diffolves iron, and is diminifhed by it as much as when it is in its native ftate of condenfation.

It is fomething remarkable, though I never attended to it particularly before I made thefe laft experiments, and it may tend to throw fome
light

light upon them, that when a candle is extinguifhed, as it never fails to be, in nitrous air,
the flame feems to be a little enlarged at its
edges, by another bluifh flame added to it, juft
before its extinction.

It is proper to obferve in this place, that the
electric fpark taken in nitrous air diminifhes it
to one fourth of its original quantity, which
is about the quantity of its diminution by iron
filings and brimftone, and alfo by liver of fulphur without heat. The air is alfo brought
by electricity to the fame ftate as it is by iron
filings and brimftone, not diminifhing common
air. If the electric fpark be taken in it when
it is confined by water tinged with archil, it is
prefently changed from blue to red, and that to
a very great degree.

When the iron nails or wires, which I have
ufed to diminifh nitrous air, had done their of-
fice, I laid them afide, not fufpecting that they
could be of any other philofophical ufe ; but
after having lain expofed to the open air almoft a fortnight ; having, for fome other purpofe, put fome of them into a veffel containing common air, ftanding inverted, and immerfed in water, I was furprized to obferve
that the air in which they were confined was
diminifhed. The diminution proceeded fo faft,

that

that the procefs was completed in about twenty-four hours ; for in that time the air was diminifhed about one fifth, fo that it made no
effervefcence with nitrous air, and was, therefore, no doubt, highly noxious, like air diminifhed by any other procefs.

This experiment I have repeated a great
number of times, with the fame phials, filled
with nails or wires that have been fuffered to
ruft in nitrous air, but their power of diminifhing common air grows lefs and lefs continually.
How long it will be before it is quite exhaufted
I cannot tell. This diminution of air I conclude muft arife from the phlogifton, either of
the nitrous air or the iron, being fome way en-
tangled in the ruft, in which the wires were en-
crufted, and afterwards getting loofe from it.

To the experiments upon iron filings and
brimftone in nitrous air, I muft add, that when
a pot full of this mixture had abforbed as much
as it could of a jar of nitrous air (which is about
three fourths of the whole) I put frefh nitrous
air to it, and it continued to abforb, till three
or four jars full of it difappeared ; but the abforption was exceedingly flow at the laft. Alfo
when I drew this pot through the water, and
admitted frefh nitrous air to it, it abforbed another

other jar full, and then ceafed. But when I fcraped off the outer furface of this mixture, which had been fo long expofed to the nitrous air, the remainder abforbed more of the air.

When I took the top of the mixture which I had fcraped off and threw upon it the focus of a burning glafs, the air in which it was confined was diminifhed, and became quite noxious; yet when I endeavoured to get air from this matter in a jar full of quickfilver, I was able to procure little or nothing.

It is not a little remarkable that nitrous air diminifhed by iron filings and brimftone, which is about one fourth, cannot, by agitation in water, be diminifhed much farther; whereas pure nitrous air may, by the fame procefs, be diminifhed to one twentieth of its whole bulk, and perhaps much more. This is fimilar to the effect of the fame mixture, and of phlogifton in other cafes, on fixed air; for it fo far changes its conftitution, that it is afterwards incapable of mixing with water. It is fimilar alfo to the effect of phlogifton in acid air, which of itfelf is almoft inftantly abforbed by water; but by this addition it is firft converted into inflammable air, which does not readily mix with water, and which, by long agitation

tion in water, becomes of another conftitution, ftill lefs mifcible with water.

I fhall clofe this fection with a few other obfervations of a mifcellaneous nature.

Nitrous air is as much diminifhed both by iron filings, and alfo by liver of fulphur, when confined in quickfilver, as when it is ex-pofed to water.

Diftilled water tinged blue with the juice of turnfole becomes red on being impregnated with nitrous air ; but by being expofed a week or a fortnight to the common atmofphere, in open and fhallow veffels, it recovers its blue colour; though, in that time, the greater part of the water will be evaporated. This fhews that in time nitrous air efcapes from the wa-ter with which it is combined, juft as fixed air does, though by no means fo readily *.

Having diffolved filver, copper, and iron in equal quantities of fpirit of nitre diluted with water, the quantities of nitrous air produced from them were in the following proportion ; from iron 8, from copper $6\frac{1}{4}$, from filver 6. In

* I have not repeated this experiment with that variation of circumftances which an attention to Mr. Bewley's obfer-vation will fuggeft.

about

about the fame proportion alfo it was neceffary to mix water with the fpirit of nitre in
each cafe, in order to make it diffolve thefe
metals with equal rapidity, filver requiring the
leaft water, and iron the moft.

Phofphorus gave no light in nitrous air,
and did not take away from its power of diminifhing common air ; only when the rednefs
of the mixture went off, the veffel in which
it was made was filled with white fumes, as
if there had been fome volatile alkali in it.
The phofphorus itfelf was unchanged.

There is fomething remarkable in the effect
of nitrous air on *infects* that are put into it.
I obferved before that this kind of air is as
noxious as any whatever, a moufe dying the
moment it is put into it ; but frogs and fnails
(and therefore, probably, other animals whofe
refpiration is not frequent) will bear being expofed to it a confiderable time, though they
die at length. A frog put into nitrous air
ftruggled much for two or three minutes,
and moved now and then for a quarter of an
hour, after which it was taken out, but did not
recover. *Wafps* always died the moment they
were put into the nitrous air. I could never
obferve that they made the leaft motion in it,
nor could they be recovered to life afterwards.

This

This was alfo the cafe in general with *fpiders, flies,* and *butterflies.* Sometimes, however, fpiders would recover after being expofed about a minute to this kind of air.

Confidering how fatal nitrous air is to infects, and likewife its great antifeptic power, I conceived that confiderable ufe might be made of it in medicine, efpecially in the form of *clyfters,* in which fixed air had been applied with fome fuccefs; and in order to try whether the bowels of an animal would bear the injection of it, I contrived, with the help of Mr. Hey, to convey a quantity of it up the anus of a dog. But he gave manifeft figns of uneafinefs, as long as he retained it, which was a confiderable time, though in a few hours afterwards he was as lively as ever, and feemed to have fuffered nothing from the operation.

Perhaps if nitrous air was diluted either with common air, or fixed air, the bowels might bear it better, and ftill it might be deftructive to *worms* of all kinds, and be of ufe to check or correct putrefaction in the inteftinal canal, or other parts of the fyftem. I repeat it once more that, being no phyfician, I run no rifk by fuch propofals as thefe; and I cannot help flattering myfelf that, in time, very great medicinal ufe will be made of the appli-

cati

cation of thefe different kinds of air to the
animal fyftem. Let ingenious phyficians attend to this fubject, and endeavour to lay hold
of the new *handle* which is now prefented them,
before it be feized by rafh empiricks ; who, by
an indifcriminate and injudicious application,
often ruin the credit of things and procefles
which might otherwife make an ufeful addition
to the *materia* and *ars medica*.

In the firft publication of my papers, having
experienced the remarkable antifeptic power of
nitrous air, I propofed an attempt to preferve
anatomical preparations, &c. by means of it;
but Mr. Hey, who made the trial, found that,
after fome months, various animal fubftances
were fhriveled, and did not preferve their natural forms in this kind of air.

S E C-

SECTION IV.

Of Marine Acid Air.

In my former experiments on this species of air I procured it from spirit of salt, but I have since hit upon a much less expensive method of getting it, by having recourse to the process by which the spirit of salt is itself originally made. For this purpose I fill a small phial with common salt, pour upon it a small quantity of concentrated oil of vitriol, and receive the fumes emitted by it in a vessel previously filled with quickfilver, and standing in a bason of quickfilver, in which it appears in the form of a perfectly *transparent air*, being precisely the same thing with that which I had before expelled from the spirit of salt.

This method of procuring acid air is the more convenient, as a phial, once prepared in this manner, will suffice, for common experiments, many weeks; especially if a little more oil of vitriol be occasionally put to it. It only requires a little more heat at the last than at the first. Indeed, at the first, the heat of a person's hand will often be sufficient to make it

throw

throw out the vapour. In warm weather it
will even keep ſmoking many days without
the application of any other heat.

On this account, it ſhould be placed where
there are no inſtruments, or any thing of metal,
that can be corroded by this acid vapour. It
is from dear-bought experience that I give this
advice. It may eaſily be perceived when this
phial is throwing out this acid vapour, as it
always appears, in the open air, in the form
of a light cloud; owing, I ſuppoſe, to the
acid attracting to itſelf, and uniting with,
the moiſture that is in the common atmoſphere.

By this proceſs I even made a ſtronger ſpirit
of ſalt than can be procured in any other way.
For having a little water in the veſſel which
contains the quickſilver, it imbibes the acid
vapour, and at length becomes truly ſaturated
with it. Having, in this manner, impregnated pure water with acid air, I could afterwards expel the ſame air from it, as from common ſpirit of ſalt.

I obſerved before that this acid vapour, or
air, has a ſtrong affinity with *phlogiſton*, ſo
that it decompoſes many ſubſtances which contain it, and with them forms a permanently inflammable air, no more liable to be imbibed
by

by water than inflammable air procured by
any other.procefs, being in fact the very fame
thing ; and that, in fome cafes, it even diflodges
fpirit of nitre and oil of vitriol, which in general appear to be ftronger acids than itfelf.
I have fince obferved that, by giving it more
time, it will extract phlogifton from fubftances
from which I at firft concluded that it was not
able to do it, as from dry wood, crufts of
bread not burnt, dry flefh, and what is more
extraordinary from flints. As there was fomething peculiar to itfelf in the procefs or refult
of each of thefe experiments, it may not be improper to mention them diftinctly.

Pieces of dry *cork wood* being put to the acid
air, a fmall quantity remained not imbibed by
water, and was inflammable.

Very dry pieces of *oak*, being expofed to
this air a day and a night, after imbibing a
confiderable quantity of it, produced air which
was inflammable indeed, but in the flighteft degree imaginable. It feemed to be very nearly
in the ftate of common air.

A piece of *ivory* imbibed the acid vapour
very flowly. In a day and a night, however,
about half an ounce meafure of permanent air
was produced, and it was pretty ftrongly in-

Q 4   flam-

flammable. The ivory was not difcoloured, but was rendered fuperficially foft, and clammy, tafting very acid.

Pieces of *beef*, roafted, and made quite dry, but not burnt, abforbed the acid vapour flowly; and when it had continued in this fituation all night, from five ounce meafures of the air, half a meafure was permanent, and pretty ftrongly inflammable. This experiment fucceeded a fecond time exactly in the fame manner; but when I ufed pieces of white dry *chicken-flefh* though I allowed the fame time, and in other refpects the procefs feemed to go on in the fame manner, I could not perceive that any part of the remaining air was inflammable.

Some pieces of a whitifh kind of *flint*, being put into a quantity of acid air, imbibed but a very little of it in a day and a night; but of $2\frac{1}{4}$ ounce meafures of it, about half a meafure remained unabforbed by water, and this was ftrongly inflammable, taking fire juft like an equal mixture of inflammable and common air. At another time, however, I could not procure any inflammable air by this means, but to what circumftance thefe different refults were owing I cannot tell.

That

That inflammable air is produced from *charcoal* in acid air I obferved before. I have fince found that it may likewife be procured from *pit coal*, without being charred.

Inflammable air I had alfo obferved to arife from the expofure of fpirit of wine, and various *oily* fubftances, to the vapour of fpirit of falt. I have fince made others of a fimilar nature, and as peculiar circumftances attended fome of thefe experiments, I fhall recite them more at large.

Effential oil of mint abforbed this air pretty faft, and prefently became of a deep brown colour. When it was taken out of this air it was of the confiftence of treacle, and funk in water, fmelling differently from what it did before ; but ftill the fmell of the mint was predominant. Very little or none of the air was fixed, fo as to become inflammable ; but more time would probably have produced this effect.

Oil of turpentine was alfo much thickened, and became of a deep brown colour, by being faturated with acid air.

Ether abforbed acid air very faft, and became firft of a turbid white, and then of a yellow

low and brown colour: In one night a confiderable quantity of permanent air was produced,
and it was ftrongly inflammable.

Having, at one time, fully fatufated a quantity of ether with acid air, I admitted bubbles
of common air to it, through the quickfilver,
by which it was confined, and obferved that
white fumes were made in it, at the entrance of
every bubble, for a confiderable time.

At another time, having fully faturated a
fmall quantity of ether with acid air, and
having left the phial in which it was contained
nearly full of the air, and inverted, it was by
fome accident overturned ; when, inftantly, the
whole room was filled with a vifible fume, like a
white cloud, which had very much the fmell of
ether, but peculiarly offenfive. Opening the
door and window of the room, this light cloud
filled a long paffage, and another room. In the
mean time the ether was feemingly all va-
nifhed, but fome time after the furface of the
quickfilver in which the experiment had been
made was covered with a liquor that tafted very
acid ; arifing, probably, from the moifture in
the atmofphere attracted by the acid vapour
with which the ether had been impregnated.

This

This viſible cloud I attribute to the union
of the moiſture in the atmoſphere with the compound of the acid air and ether. I have
ſince ſaturated other quantities of ether with
acid air, and found it to be exceedingly volatile,
and inflammable. Its exhalation was alſo viſible, but not in ſo great a degree as in the caſe
above mentioned.

Camphor was preſently reduced into a fluid
ſtate by imbibing acid air, but there ſeemed to
be ſomething of a whitiſh ſediment in it. After
continuing two days in this ſituation I admitted
water to it; immediately upon which the camphor reſumed its former ſolid ſtate, and, to appearance, was the very ſame ſubſtance that it
had been before; but the taſte of it was acid,
and a very ſmall part of the air was permanent,
and ſlightly inflammable.

The acid air ſeemed to make no impreſ-
ſion upon a piece of Derbyſhire *ſpar*, of a very
dark colour, and which, therefore, ſeemed to
contain a good deal of phlogiſton.

As the acid air has ſo near an affinity with
phlogiſton, I expected that the fumes of *liver of
ſulphur*, which chemiſts agree to be phlogiſtic,
would have united with it, ſo as to form inflammable air; but I was diſappointed in that ex-
pecta-

pectation. This fubftance imbibed half of the
acid air to which it was introduced : one fourth
of the remainder, after ftanding one day in
quickfilver, was imbibed by water, and what
was left extinguifhed a candle. This experiment, however, feems to prove that acid air
and phlogifton may form a permanent kind of
air that is not inflammable. Perhaps it may be
air in fuch a ftate as common air loaded with
phlogifton, and from which the fixed air has
been precipitated. Or rather, it may be the
fame thing with inflammable air, that has loft
its inflammability by long ftanding in water.
It well deferves a farther examination.

The following experiments are thofe in which
the *ftronger acids* were made ufe of, and therefore they may affift us farther to afcertain their
affinities with certain fubftances, with refpect to
this marine acid in the form of air.

I put a quantity of ftrong concentrated *oil of
vitriol* to acid air, but it was not at all affected
by it in a day and a night. In order to try
whether it would not have more power in a
more condenfed ftate, I compreffed it with an
additional atmofphere; but upon taking off this
preffure, the air expanded again, and appeared
to be not at all diminifhed. I alfo put a quantity of ftrong *fpirit of nitre* to it without any
fen-

fenfible effect. We may conclude, therefore, that the marine acid, in this form of air, is not able to diflodge the other acids from their union with water.

Blue vitriol, which is formed by the union of the vitriolic acid with copper, turned to a dark green the moment that it was put to the acid air, which it abforbed, though flowly. Two pieces, as big as fmall nuts, abforbed three ounce meafures of the air in about half an hour. The green colour was very fuperficial ; for it was eafily wiped or wafhed off.

Green copperas turned to a deeper green upon being put into acid air, which it abforbed flowly. *White copperas* abforbed this air very faft, and was diffolved in it.

Sal ammoniac, being the union of fpirit of falt with volatile alkali, was no more affected with the acid air than, as I have obferved before, common falt was.

I alfo introduced to the acid air various other fubftances, without any particular expectation ; and it may be worth while to give an account of the refults, that the reader may draw from them fuch conclufions as he fhall think reafonable.

Borax

Borax abforbed acid air about as faft as blue vitriol, but without any thing elfe that was obfervable.

Fine white *fugar* abforbed this air flowly, was thoroughly penetrated with it, became of a deep brown colour, and acquired a fmell that was peculiarly pungent.

A piece of *quick lime* being put to about twelve or fourteen ounce meafures of acid air, and continuing in that fituation about two days, there remained one ounce meafure of air that was not abforbed by water, and it was very ftrongly inflammable, as much fo as a mixture of half inflammable and half common air. Very particular care was taken that no common air mixed with the acid air in this procefs. At another time, from about half the quantity of acid air above mentioned, with much lefs quicklime, and in the fpace of one day, I got half an ounce meafure of air that was inflammable in a flight degree only. This experiment proves that fome part of the phlogifton which efcapes from the fuel, in contact with which the lime is burned, adheres to it. But I am very far from thinking that the caufticity of quick lime is at all owing to this circumftance.

I have

I have made a few more experiments on tne mixture of acid air with *other kinds of air* and think that it may be worth while to mention them, though nothing of confequence, at leaft nothing but negative conclufions, can be drawn from them.

A quantity of common air faturated with nitrous air was put to a quantity of acid air, and they continued together all night, without any fenfible effect. The quantity of both re- mained the fame, and water being admitted to them, it abforbed all the acid air, and left the other juft as before.

A mixture of two thirds of air diminifhed by iron filings and brimftone, and one third acid air, were mixed together, and left to ftand four weeks in quickfilver. But when the mix- ture was examined, water prefently imbibed all the acid air, and the diminifhed air was found to be juft the fame that it was before. I had ima- gined that the acid air might have united with the phlogifton with which the diminifhed air was overcharged, fo as to render it wholfome ; and I had read an account of the ftench arifing from putrid bodies being corrected by acid fumes.

The

The remaining experiments, in which the acid air was principally concerned, are of a mifcellaneous nature.

I put a piece of dry *ice* to a quantity of acid air (as was obferved in the fection concerning *alkaline* air) taking it with a forceps, which, as well as the air itfelf, and the quickfilver by which it had been confined, had been expofed to the open air for an hour, in a pretty ftrong froft. The moment it touched the air it was diffolved as faft as it would have been by being thrown into a hot fire, and the air was prefently imbibed. Putting frefh pieces of ice to that which was diffolved before, they were alfo diffolved immediately, and the water thus procured did not freeze again, though it was expofed a whole night, in a very intenfe froft.

Flies and fpiders die in acid air, but not fo quickly as in nitrous air. This furprized me very much ; as I had imagined that nothing could be more fpeedily fatal to all animal life than this pure acid vapour.

As inflammable air, I have obferved, fires at one explofion in the vapour of fmoking fpirit of nitre, juft like an equal mixture of inflammable and common air, I thought it was poffible that the fume which naturally rifes from common

mon fpirit of falt might have the fame effect, but it had not. For this purpofe I treated the fpirit of falt, as I had before done the fmoking fpirit of nitre ; firft filling a phial with it, then inverting it in a veffel containing a quantity of the fame acid ; and having thrown the .inflammable air into it, and thereby driven out all the acid, turning it with its mouth upwards, and immediately applying a candle to it.

Acid air not being fo manageable as moft of the other kinds of air, I had recourfe to the following peculiar method, in order to afcertain its *fpecific gravity.* Having filled an eight ounce phial with this air, and corked it up, I weighed it very accurately ; and then, taking out the cork, I blew very ftrongly into it with a pair of bellows, that the common air might take place of the acid ; and after this I weighed it again, together with the cork, but I could not perceive the leaft difference in the weight. I conclude, however, from this experiment, that the acid air is heavier than the common air, becaufe the mouth of the phial and the infide of it were evidently moiftened by the water which the acid vapour had attracted from the air, which moifture muft have added to the weight of the phial.

R   SECTION

SECTION V.

Of INFLAMMABLE AIR.

It will have appeared from my former experiments, that inflammable air confifts chiefly,
if not wholly, of the union of an acid vapour
with phlogifton; that as much of the phlogifton
as contributes to make air inflammable is imbibed by the water in which it is agitated; that
in this procefs it foon becomes fit for refpiration,
and by the continuance of it comes at length
to extinguifh flame. Thefe obfervations, and
others which I have made upon this kind of
air, have been confirmed by my later experiments, efpecially thofe in which I have connected *electrical experiments* with thofe on air.

The electric fpark taken in any kind of *oil*
produces inflammable air, as I was led to obferve in the following manner. Having
found, as will be mentioned hereafter, that
ether doubles the quantity of any kind of air
to which it is admitted; and being at that time
engaged in a courfe of experiments to afcertain
the effect of the electric matter on all the different kinds of air, I had the curiofity to try
what it would do with *common air*, thus increafed

creafed by means of ether. The very firft fpark, I obferved, increafed the quantity of this air very confiderably, fo that I had very foon fix or eight times as much as I began with; and whereas water imbibes all the ether that is put to any kind of air, and leaves it without any vifible change, with refpect to quantity or quality, this air, on the contrary, was not imbibed by water. It was alfo very little diminifhed by the mixture of nitrous air From whence it was evident, that it had received an addition of fome other kind of air, of which it now principally confifted.

In order to determine whether this effect was produced by the *wire*, or the *cement* by which the air was confined (as I thought it poffible that phlogifton might be difcharged from them) I made the experiment in a glafs fyphon, fig. 19, and by that means I contrived to make the electric park pafs from quickfilver through the air on which I made the experiment, and the effect was the fame as before. At one time there happened to be a bubble of common air, without any ether, in one part of the fyphon, and another bubble with ether in another part of it; and it was very amufing to obferve how the fame electric fparks diminifhing the former of thefe bubbles, and increafed the latter.

It

It being evident that the *ether* occafioned the difference that was obfervable in thefe two cafes, I next proceeded to take the electric fpark in a quantity of ether only, without any air whatever; and obferved that every fpark produced a fmall bubble; and though, while the fparks were taken in the ether itfelf, the generation of air was flow, yet when fo much air was collected, that the fparks were obliged to pafs through it, in order to come to the ether and the quickfilver on which it refted, the increafe was exceedingly rapid; fo that, making the experiment in fmall tubes, as fig. 16, the quickfilver foon receded beyond the ftriking diftance. This air, by paffing through water, was diminifhed to about one third, and was inflammable.

One quantity of air produced in this manner from ether I fuffered to ftand two days in water, and after that I transferred it feveral times through the water, from one veffel to another, and ftill found that it was very ftrongly inflammable; fo that I have no doubt of its being genuine inflammable air, like that which is produced from metals by acids, or by any other chemical procefs.

Air produced from ether, mixed both with common and nitrous air, was likewife inflammable;

mable; but in the cafe of the nitrous air, the original quantity bore a very fmall proportion to the quantity generated.

Concluding that the inflammable matter in this air came from the ether, as being of the clafs of *oils*, I tried other kinds of oil, as *oil of olives, oil of turpentine,* and *effential oil of mint,* taking the electric fpark in them, without any air to begin with, and found that inflammable air was produced in this manner from them all. The generation of air from oil of turpentine was the quickeft, and from the oil of olives the floweft in thefe three cafes.

By the fame procefs I got inflammable air from *fpirit of wine,* and about as copioufly as from the effential oil of mint. This air continued in water a whole night, and when it was transferred into another veffel was ftrongly inflammable.

In all thefe cafes the inflammable matter might be fuppofed to arife from the inflammable fubftances on which the experiments were made. But finding that, by the fame procefs, I could get inflammable air from the *volatile fpirit of fal ammoniac,* I conclude that the phlogifton was in part fupplied by the electric matter itfelf. For though, as I have obferved be-

R 3   fore,

fore, the alkaline air which is expelled from
the fpirit of fal ammoniac be inflammable, it
is fo in a very flight degree, and can only be
perceived to be fo when there is a confiderable
quantity of it.

Endeavouring to procure air from a cauftic
alkaline liquor, accurately made for me by Mr.
Lane, and alfo from fpirit of falt, I found
that the electric fpark could not be made vifible in either of them ; fo that they muft be
much more perfect conductors of electricity
than water, or other fluid fubftances. This
experiment well deferves to be profecuted.

I obferved before that inflammable air, by
ftanding long in water, and efpecially by agitation in water, lofes its inflammability ; and that
in the latter cafe, after paffing through a ftate
in which it makes fome approach to common
air (juft admitting a candle to burn in it) it
comes to extinguifh a candle. I have fince made
another obfervation of this kind, which well
deferves to be recited. It relates to the inflammable air generated from oak the 27th of July
1771, of which I have made mention before.

This air I have obferved to have been but
weakly.inflammable fome months after it was
generated, and to have been converted into
pretty

pretty good or wholefome air by no great degree of agitation in water ; but on the 27th of
March 1773, I found the remainder of it to
be exceedingly good air. A candle burned in
it perfectly well, and it was diminifhed by nitrous air almoft as much as common air.

I fhall conclude this fection with a few mifcellaneous obfervations of no great importance.

Inflammable air is not changed by being
made to pafs many times through a red hot iron
tube. It is alfo no more diminifhed or changed by the fumes of liver of fulphur, or by the
electric fpark, than I have before obferved it to
have been by a mixture of iron filings and
brimftone. When the electric fpark was taken
in it, it was confined by a quantity of water
tinged blue with the juice of archil, but the
colour remained unchanged.

I put two *wafps* into inflammable air, and
let them remain there a confiderable time, one
of them near an hour. They prefently ceafed
to move, and feemed to be quite dead for about
half an hour after they were taken into the
open air ; but then they came to life again, and
prefently after feemed to be as well as ever they
had been.

R 4   SEC-

SECTION VI.

Of FIXED AIR.

The additions I have made to my obfervations on *fixed air* are neither numerous nor
confiderable.

The moſt important of them is a confirmation of my conjecture, that fixed air is capable
of forming an union with phlogifton, and thereby becoming a kind of air that is not mifcible
with water. I had produced this effect before
by means of iron filings and brimftone, fermenting in this kind of air ; but I have fince
had a much more decifive and elegant proof of
it by *electricity*. For after taking a fmall electric explofion, for about an hour, in the fpace
of an inch of fixed air, confined in a glafs tube
one tenth of an inch in diameter, fig. 16, I
found that when water was admitted to it, only
one fourth of the air was imbibed. Probably
the whole of it would have been rendered immifcible in water, if the electrical operation had
been continued a fufficient time. This air continued feveral days in water, and was even agitated in water without any farther diminution.

It

It was not, however, common air, for it was not diminifhed by nitrous air.

By means of iron filings and brimftone I have, fince my former experiments, procured a confiderable quantity of this kind of air in a method fomething different from that which I ufed before. For having placed a pot of this mixture under a receiver, and exhaufted it with a pump of Mr. Smeaton's conftruction, I filled it with fixed air, and then left it plunged under water; fo that no common air could have accefs to it. In this manner, and in about a week, there was, as near as I can recollect, one fixth, or at leaft one eighth of the whole converted into a permanent air, not imbibed by water.

From this experiment I expected that the fame effect would have been produced on fixed air by the fumes of *liver of fulphur*; but I was difappointed in that expectation, which furprifed me not a little; though this correfponds in fome meafure, to the effect of phlogifton exhaled from this fubftance on acid air. Perhaps more time may be requifite for this purpofe, for this procefs was not continued more than a day and a night.

Iron

Iron filings and brimstone, I have observed, ferment with great heat in nitrous air, and I have since observed that this procefs is attended with greater heat in fixed air than in common air.

Though fixed air incorporated with water diffolves iron, fixed air without water has no fuch power, as I obferved before. I imagined that, if it could have diffolved iron, the phlogifton would have united with the air, and have made it immifcible with water, as in the former inftances; but after being confined in a phial full of nails from the 15th of December to the 4th of October following, neither the iron nor the air appeared to have been affected by their mutual contact.

Having expofed equal quantities of common and fixed air, in equal and fimilar cylindrical glafs veffels, to equal degrees of heat, by placing them before a fire, and frequently changing their fituations, I obferved that they were expanded exactly alike, and when removed from the fire they both recovered their former dimenfions.

Having had fome fmall fufpicion that liver of fulphur, befides emitting phlogifton, might alfo

alfo yield fome fixed air (which is known to be contained in the falt of tartar from which it is made) I mixed the two ingredients, viz. falt of tartar and brimftone, and putting them into a thin phial, and applying the flame of a candle to it, fo as to form the liver of fulphur, I received the air that came from it in this procefs in a veffel of quickfilver. In this manner I procured a very confiderable quantity of fixed air, fo that I judged it was all difcharged from the tartar. But though it is poffible that a fmall quantity of it may remain in liver of fulphur, when it is made in the moft perfect manner, it is not probable that it can be expelled without heat.

S E C-

SECTION VII.

MISCELLANEOUS EXPERIMENTS,

1. It is fomething extraordinary that, though
ether, as I found, cannot be made to affume
the form of air (the vapour arifing from it by
heat, being foon condenfed by cold, even in
quickfilver) yet that a very fmall quantity of
ether put to any kind of air, except the acid,
and alkaline, which it imbibes, almoft inftantly doubles the apparent quantity of it ; but
upon paffing this air through water, it is
prefently reduced to its original quantity again,
with little or no change of quality.

I put about the quantity of half a nut-fhell
full of ether, inclofed in a glafs tube, through
a body of quickfilver, into an ounce meafure
of common air, confined by quickfilver ; upon
which it prefently began to expand, till it occupied the fpace of two ounce meafures. It
then gradually contracted about one fixth of an
ounce meafure. Putting more ether to it, it
again expanded to two ounce meafures ; but no
more addition of ether would make it expand
any farther. Withdrawing the quickfilver, and
admitting water to this air, without any agitation,

tion, it began to be abforbed; but only about half an ounce meafure had difappeared after it had ftood an hour in the water. But by once paffing it through water the air was reduced to its original dimenfions. Being tried by a mixture of nitrous air, it appeared not to be fo good as frefh air, though the injury it had received was not confiderable.

All the phenomena of dilatation and contraction were nearly the fame, when, inftead of common air, I ufed nitrous air, fixed air, inflammable air, or any fpecies of phlogifticated common air. The quantity of each of thefe kinds of air was nearly doubled while they were kept in quickfilver, but fixed air was not fo much increafed as the reft, and phlogifticated air lefs; but after paffing through the water, they appeared not to have been fenfibly changed by the procefs.

2. Spirit of wine yields no air by means of heat, the vapours being foon condenfed by cold, like the vapour of water; yet when, in endeavouring to procure air from it, I made it boil, and catched the air which had refted on the furface of the fpirit, and which had been expelled by the heat together with the vapour, in a veffel of quickfilver, and afterwards admitted acid air to it, the veffel was filled with white fumes,

as

as if there had been a mixture of alkaline air along with it. To what this appearance was owing I cannot tell, and indeed I did not examine into it.

3. Having been informed by Dr. Small and Mr. Bolton of Birmingham, that paper dipped in a folution of copper in fpirit of nitre would take fire with a moderate heat (a fact which I afterwards found mentioned in the Philofophical Tranfactions) it occurred to me that this would be very convenient for experiments relating to *ignition* in different kinds of air ; and indeed I found that it was eafily fired, either by a burning lens, or the approach of red-hot iron on the outfide of the phial in which it was contained, and that any part of it being once fired, the whole was prefently reduced to afhes ; provided it was previoufly made thoroughly dry, which, however, it is not very eafy to do.

With this preparation, I found that this paper burned freely in all kinds of air, but not in *vacuo*, which is alfo the cafe with gunpowder ; and, as I have in effect obferved before, all the kinds of air in which this paper was burned received an addition to their bulk, which confifted partly of nitrous air, from the nitrous precipitate, and partly of inflammable air, from the paper. As fome of the circumftances

ftances attending the ignition of this paper in fome of the kinds of air were a little remarkable, I fhall juft recite them.

Firing this paper in *inflammable* air, which it did without any ignition of the inflammable air itfelf, the quantity increafed regularly, till the phial in which the procefs was made was nearly full; but then it began to decreafe, till one third of the whole quantity difappeared.

A piece of this paper being put to three ounce meafures of *acid* air, a great part of it prefently turned yellow, and the air was reduced to one third of the original quantity, at the fame time becoming reddifh, exactly like common air in a phial containing fmoking fpirit of nitre. After this, by the approach of hot iron, I fet fire to the paper; immediately upon which there was a production of air which more than filled the phial. This air appeared, upon examination, to be very little different from pure nitrous air. I repeated this experiment with the fame event.

Paper dipped in a folution of mercury, zinc, or iron, in nitrous acid, has, in a fmall degree, the fame property with paper dipped in a folution of copper in the fame acid.

Gun.

4. Gunpowder is alfo fired in all kinds of air, and, in the quantity in which I tried it, did not make any fenfible change in them, except that the common air in which it was fired would not afterwards admit a candle to burn in it. In order to try this experiment I half ex-haufted a receiver, and then with a burning-glafs fired the gunpowder which had been pre-vioufly put into it. By this means I could fire a greater quantity of gunpowder in a fmall quan-tity of air, and avoid the hazard of blowing up, and breaking my receiver.

I own that I was rather afraid of firing gunpowder in inflammable air, but there was no reafon for my fear ; for it exploded quite freely in this air, leaving it, in all refpects, juft as it was before.

In order to make this experiment, and in-deed almoft all the experiments of firing gun-powder in different kinds of air, I placed the powder upon a convenient ftand within my re-ceiver, and having carefully exhaufted it by a pump of Mr. Smeaton's conftruction, I filled the receiver with any kind of air by the appa-ratus defcribed, p. 19, fig. 14, taking the great-eft care that the tubes, &c. which conveyed the air fhould contain little or no common air. In the experiment with inflammable air a
con-

confiderable mixture of common air would have been exceedingly hazardous : for, by that affift-ance, the inflammable air might have exploded in fuch a manner, as to have been dangerous to the operator. Indeed, I believe I fhould not have ventured to have made the experiment at all with any other pump befides Mr. Smeaton's.

Sometimes, I filled a glafs veffel with quick-filver, and introduced the air to it, when it was inverted in a bafon of quickfilver. By this means I intirely avoided any mixture of common air ; but then it was not eafy to convey the gunpow-der into it, in the exact quantity that was re-quifite for my purpofe. This, however, was the only method by which I could contrive to fire gunpowder in acid or alkaline air, in which it exploded juft as it did in nitrous or fixed air.

I burned a confiderable quantity of gunpow-der in an exhaufted receiver (for it is well-known that it will not explode in it) but the air I got from it was very inconfiderable, and in thefe circumftances was neceffarily mixed with common air. A candle would not burn in it.

S

S E C-

SECTION VIII.

QUERIES, SPECULATIONS, *and* HINTS.

I begin to be apprehenfive leſt, after being confidered as a *dry experimenter*, I ſhould paſs, with many of my readers, into the oppoſite character of a *viſionary theoriſt*. A good deal of theory has been interfperfed in the courſe of this work, but, not content with this, I am now entering upon a long ſection, which contains nothing elſe.

The conjectures that I have ventured to advance in the body of the work will, I hope, be found to be pretty well fupported by facts ; but the prefent ſection will, I acknowledge, contain many *random thoughts*. I have, however, thrown them together by themfelves, that readers of lefs imagination, and who care not to advance beyond the regions of plain fact, may, if they pleaſe, proceed no farther, that their delicacy be not offended.

In extenuation of my offence, let it, however, be confidered, that *theory* and *experiment*

ne-

neceffarily go hand in hand, every procefs being intended to afcertain fome particular *hypothefis*, which, in fact, is only a conjecture concerning the circumftances or the caufe of
fome natural operation; confequently that the
boldeft and moft original experimenters are
thofe, who, giving free fcope to their imagina-
tions, admit the combination of the moft diftant
ideas; and that though many of thefe affocia-
tions of ideas, will be wild and chimerical, yet
that others will have the chance of giving rife
to the greateft and moft capital difcoveries;
fuch as very cautious, timid, fober, and flow-
thinking people would never have come at.

Sir Ifaac Newton himfelf, notwithftanding
the great advantage which he derived from a
habit of *patient thinking*, indulged bold and
excentric thoughts, of which his Queries at
the end of his book of Optics are a fufficient
evidence. And a quick conception of diftant
analogies, which is the great key to unlock the
fecrets of nature, is by no means incompatible
with the fpirit of *perfeverance*, in inveftigations
calculated to afcertain and purfue thofe analogies.

§ 1. *Speculations concerning the* CONSTITUENT
PRINCIPLES *of the different kinds of* AIR, *and
the* CONSTITUTION *and* ORIGIN *of the* ATMO-
SPHERE, &c.

All the kinds of air that appear to me to be
effentially diftinct from each other are *fixed
air, acid* and *alkaline* ; for thefe, and another
principle, called *phlogifton,* which I have not
been able to exhibit in the form of *air,* and
which has never yet been exhibited by itfelf in
any form, feem to conftitute all the kinds of air
that I am acquainted with.

Acid air and phlogifton conftitute an air
which either extinguifhes flame, or is itfelf inflammable, according, probably, to the quantity
of phlogifton combined in it, or the mode of combination. When it extinguifhes flame, it is probably only fo much charged with the phlogiftic
matter, as to take no more from a burning candle, which muft, therefore, neceffarily go out
in it. When it is inflammable, it is probably
fo much charged with phlogifton, that the heat
communicated by a burning candle makes it
immediately feparate itfelf from the other principle with which it was united, in which feparation *heat* is produced, as in other cafes of ignition; the action and reaction, which neceffarily
attends the feparation of the conftituent principles,

ples, exciting probably a vibratory motion in them.

Since inflammable air, by agitation in water, firft comes to lofe its inflammability, fo as to be fit for refpiration, and even to admit a candle to burn in it, and then comes to extinguifh a candle ; it feems probable that water imbibes a great part of this extraordinary charge of phlogifton. And that water *can* be impregnated with phlogifton, is evident from many of my experiments, efpecially thofe in which metals were calcined over it.

Water having this affinity with phlogifton, it is probable that it always contains a confider-able portion of it ; which phlogifton having a ftronger affinity with the acid air, which is per-haps the bafis of common air, may by long agitation be communicated to it, fo as to leave it over faturated, in confequence of which it will extinguifh a candle.

It is poffible, however, that inflammable air and air which extinguifhes a candle may differ from one another in the *mode* of the combina-tion of thefe two conftituent principles, as well as in the proportional quantity of each ; and by agitation in water, or long ftanding, that mode of combination may change. This we

S 3   know

know to be the cafe with other fubftances, as
with *milk*, from which, by ftanding only, *cream*
is feparated ; which by agitation becomes *but-
ter*. Alfo many fubftances, being at reft, putrefy, and thereby become quite different from
what they were before. If this be the cafe with
inflammable air, the water may imbibe either
of the conftituent parts, whenever any proportion of it is fpontaneoufly feparated from the
reft; and fhould this ever be that phlogifton,
with which air is but flightly over-charged, as
by the burning of a candle, it will be recovered
to a ftate in which a candle may burn in it again.

It will be obferved, however, that it was only
in one inftance that I found that ftrong inflammable air, in its tranfition to a ftate in which it
extinguifhes a candle, would admit a candle to
burn in it, and that was very faintly ; that then
the air was far from being pure, as appeared by
the teft of nitrous air; and that it was only from
a particular quantity of inflammable air which
I got from oak, and which had ftood a long
time in water, that I ever got air which was as
pure as common air. Indeed, it is much more
eafy to account for the paffing of inflammable
air into a ftate in which it extinguifhes candles,
without any intermediate ftate, in which it will
admit a candle to burn in it, than otherwife.
This fubject requires and deferves farther inveftigation.

tigation. It will alfo be well worth while to examine what difference the agitation of air in
natural or artificial *fea-water* will occafion.

Since acid air and phlogifton make inflammable air, and fince inflammable air is con-
vertible into air fit for refpiration, it feems not
to be improbable, that thefe two ingredients are
the only effential principles of common air.
For this change is produced by agitation in water only, without the addition of any fixed air,
though this kind of air, like various other things
of a foreign nature, may be combined with it.

Confidering alfo what prodigious quantities
of inflammable air are produced by the burning
of fmall pieces of wood or pit-coal, it may
not be improbable but that the *volcanos*, with
which there are evident traces of almoft the
whole furface of the earth having been overfpread, may have been the origin of our atmofphere, as well as (according to the opinion
of fome) of all the folid land.

The fuperfluous phlogifton of the air, in the
ftate in which it iffues from volcanos, may have
been imbibed by the waters of the fea, which
it is probable originally covered the furface of
the earth, though part of it might have united
with the acid vapour exhaled from the fea, and

S 4 by

by this union have made a confiderable and valuable addition to the common mafs of air;
and the remainder of this over-charge of phlogifton may have been imbibed by plants as foon
as the earth was furnifhed with them.

That an acid vapour is really exhaled from
the fea, by the heat of the fun, feems to be
evident from the remarkably different ftates of
the atmofphere, in this refpect, in hot and cold
climates. In Hudfon's bay, and alfo in Ruffia,
it is faid, that metals hardly ever ruft, whereas
they are remarkably liable to ruft in Barbadoes,
and other iflands between the tropics. See
Ellis's Voyage, p. 288. This is alfo the cafe in
places abounding with falt-fprings, as Nant-
wich in Chefhire.

That mild air fhould confift of parts of fo
very different a nature as an acid vapour and
phlogifton, one of which is fo exceedingly cor-
rofive, will not appear furprifing to a chemift,
who confiders the very ftrong affinity which
thefe two principles are known to have with
each other, and the exceedingly different properties which fubftances compofed by them poffefs. This is exemplified in common *fulphur*,
which is as mild as air, and may be taken into
the ftomach with the utmoft fafety, though nothing can be more deftructive than one of its
con-

conftituent parts, feparately taken, viz. oil of vitriol. Common air, therefore, notwithftanding its mildnefs, may be compofed of fimilar principles, and be a real *fulphur.*

That the fixed air which makes part of the atmofphere is not prefently imbibed by the waters of the fea, on which it refts, may be owing to the union which this kind of air alfo appears to be capable of forming with phlogifton. For fixed air is evidently of the nature of an acid ; and it appears, in fact, to be capable of being combined with phlogifton, and thereby of conftituting a fpecies of air not liable to be imbibed by water. Phlogifton, however, having a ftronger affinity with acid air, which I fuppofe to be the bafis of common air, it is not furprifing that, uniting with this, in preference to the fixed air, the latter fhould be precipitated, whenever a quantity of common air is made noxious by an over-charge of phlogifton.

The fixed air with which our atmofphere abounds may alfo be fupplied by volcanos, from the vaft maffes of calcareous matter lodged in the earth, together with inflammable air. Alfo a part of it may be fupplied from the fermentation of vegetables upon the furface of it. At pre-

present, as fast as it is precipitated and imbibed by one procefs, it may be fet loofe by others.

Whether there be, upon the whole, an increafe or a decreafe of the general mafs of the atmofphere is not eafy to conjecture, but I fhould imagine that it rather increafes. It is true that many proceffes contribute to a great vifible diminution of common air, and that when by other proceffes it is reftored to its former wholefomenefs, it is not increafed in its dimenfions ; but volcanos and fires ftill fupply vaft quantities of air, though in a ftate not yet fit for refpiration ; and it will have been feen in my experiments, that vegetable and animal fubftances, diffolved by putrefaction, not only emit phlogifton, but likewife yield a confiderable quantity of permanent elaftic air, overloaded indeed with phlogifton, as might be expected, but capable of being purified by thofe proceffes in nature by which other noxious air is purified.

That particles are continually detaching themfelves from the furfaces of all folid bodies, even the metallic ones, and that thefe particles conftitute the moft permanent part of the atmofphere, as Sir Ifaac Newton fuppofed, does not appear to me to be at all probable.

My

My readers will have obferved, that not only is common air liable to be diminifhed by a mixture of nitrous air, but likewife air originally produced from inflammable air, and even from nitrous air itfelf, which never contained any fixed air. From this it may be inferred, that the whole of the diminution of common air by phlogifton is not owing to the precipitation of fixed air, but from a real contraction of its dimenfions, in confequence of its union with phlogifton. Perhaps an accurate attention to the fpecific gravity of air procured from thefe different materials, and in thefe different ftates, may determine this matter, and affift us in inveftigating the nature of phlogifton.

In what *manner* air is diminifhed by phlogifton, independent of the precipitation of any of its conftituent parts, is not eafy to conceive ; unlefs air thus diminifhed be heavier than air not diminifhed, which I did not find to be the cafe. It deferves, however, to be tried with more attention. That phlogifton fhould communicate abfolute *levity* to the bodies with which it is combined, is a fuppofition that I am not willing to have recourfe to, though it would afford an eafy folution of this difficulty.

I have likewife obferved, that a moufe will live almoft as long in inflammable air, when it
has

has been agitated in water, and even before it has been deprived of all its inflammability, as in common air; and yet that in this ſtate it is not, perhaps, ſo much diminiſhed by nitrous air as common air is. In this caſe, therefore, the diminution ſeems to have been occaſioned by a contraction of dimenſions, and not by a loſs of any conſtituent part; ſo that the air is really better, that is, more fit for reſpiration, than, by the teſt of nitrous air, it would ſeem to be.

If this be the caſe (for it is not eaſy to judge with accuracy by experiments with ſmall animals) nitrous air will be an accurate teſt of the goodneſs of *common air* only, that is, air containing a conſiderable proportion of fixed air. But this is the moſt valuable purpoſe for which a teſt of the goodneſs of air can be wanted. It will ſtill, indeed, ſerve for a meaſure of the goodneſs of air that does not contain fixed air; but, a ſmaller degree of diminution in this caſe, muſt be admitted to be equivalent to a greater diminution in the other.

As I could never, by means of growing vegetables, bring air which had been thoroughly noxious to ſo pure a ſtate as that a candle would burn in it, it may be ſuſpected that ſomething elſe beſides *vegetation* is neceſſary to produce this effect. But it ſhould be conſidered, that
no

no part of the common atmofphere can ever be
in this highly noxious ftate, or indeed in a ftate
in which a candle will not burn in it; but that
even air reduced to this ftate, either by candles
actually burning out in it, or by breathing it,
has never failed to be perfectly reftored by
vegetation, at leaft fo far that candles would
burn in it again, and, to all appearance,. as
well, and as long as ever; fo that the growing
vegetables, with which the furface of the earth
is overfpread, may, for any thing that appears
to the contrary, be a caufe of the purification
of the atmofphere fufficiently adequate to the
effect.

It may likewife be fufpected, that fince *agitation in water* injures pure common air, the
agitation of the fea may do more harm than
good in this refpect. But it requires a much
more violent and longer continued agitation of
air in water than is ever occafioned by the waves
of the fea to do the leaft fenfible injury to it.
Indeed a flight agitation of air in *putrid water*
injures it very materially; but if the water be
fweet, this effect is not produced, except by a
long and tedious operation, whereas it requires
but a very fhort time, in comparifon, to reftore
a quantity of any of the moft noxious kinds of
air to a very great degree of wholefomenefs by
the fame procefs.

Dr.

Dr. Hales found that he could breathe the fame air much longer when, in the courfe of his refpiration, it was made to pafs through feveral folds of cloth dipped in vinegar, in a folution of fea falt, or in falt of tartar, efpecially the laft. Statical Effays, vol. 1. p. 266. The experiment is valuable, and well deferves to be repeated with a greater variety of circumftances. I imagine that the effect was produced by thofe fubftances, or by the *water* which they attracted from the air, imbibing the phlogiftic matter difcharged from the lungs. Perhaps the phlogifton may unite with the watery part of the atmofphere, in preference to any other part of it, and may by that means be more eafily transferred to fuch falts as imbibe moifture.

Sir Ifaac Newton defines *flame* to be *fumus candens,* confidering all *fmoke* as being of the fame nature, and capable of ignition. But the fmoke of common fuel confifts of two very different things. That which rifes firft is mere *water,* loaded with fome of the groffer parts of the fuel, and is hardly more capable of becoming red hot than water itfelf ; but the other kind of fmoke, which alone is capable of ignition, is properly *inflammable air,* which is alfo loaded with other heterogeneous matter, fo as to appear like a very denfe fmoke. A lighted candle foon fhews them to be effentially different

rent from each other. For one of them inftantly takes fire, whereas the other extinguifhes a candle.

It is remarkable that gunpowder will take fire, and explode in all kinds of air, without diftinction, and that other fubftanees which contain *nitre* will burn freely in thofe circumftances. Now fince nothing can burn, unlefs there be fomething at hand to receive the phlogifton, which is fet loofe in the act of ignition, I do not fee how this fact can be accounted for, but by fuppofing that the acid of nitre, being peculiarly formed to unite with phlogifton, immediately receives it. And if the fulphur, which is thereby formed, be inftantly decompofed again, as the chemifts in general fay, thence comes the explofion of gunpowder, which, however, requires the reaction of fome incumbent atmofphere, and without which the materials will only *melt,* and be *difperfed* without explofion.

Nitrous air feems to confift of the nitrous acid vapour united to phlogifton, together, perhaps, with fome fmall portion of the metallic calx ; juft as inflammable air confifts of the vitriolic or marine acid, and the fame phlogiftic principle. It fhould feem, however, that phlogifton has a ftronger affinity with the marine acid,

acid, if that be the bafis of common air; for nitrous air being admitted to common air, it is immediately decompofed; probably by the phlogifton joining with the acid principle of the common air, while the fixed air which it contained is precipitated, and the acid of the nitrous air is abforbed by the water in which the mixture is made, or unites with any volatile alkali that happens to be at hand.

This, indeed, is hardly agreeable to the hypothefis of moft chemifts, who fuppofe that the nitrous acid is ftronger than the marine, fo as to be capable of diflodging it from any bafe with which it may be combined; but it agrees with my own experiments on marine acid air, which fhew that, in many cafes, this *weaker acid,* as it is called, is capable of feparating both the vitriolic and the nitrous acids from the phlogifton with which they are combined.

On the other hand, the folution of metals in the different acids feem to fhew, that the nitrous acid forms a clofer union with phlogifton than the other two; becaufe the air which is formed by the nitrous acid is not inflammable, like that which is produced by the oil of vitriol, or the fpirit of falt. Alfo, the fame weight of iron does not yield half the quantity of nitrous air that it does of inflammable.

The

The great diminution of nitrous air by phlogifton is not eafily accounted for, unlefs we
fuppofe that its fuperabundant acid, uniting
more intimately with the phlogifton, conftitutes
a fpecies of *fulphur* that is not eafily perceived
or catched ; though, in the procefs with iron,
and alfo in that with liver of fulphur, part of
the redundant phlogifton forms fuch an union
with the acid as gives it a kind of inflammability.

It appears to me to be very probable, that
the fpirit of nitre might be exhibited in the
form of *air*, if it were poffible to find any fluid
by which it could be confined ; but it unites
with quickfilver as well as with water, fo that
when, by boiling the fpirit of nitre, the fumes
are driven through the glafs tube, fig. 8, they
inftantly feize upon the quickfilver through
which they are to be conveyed, and uniting
with it, form a fubftance that ftops up the
tube : a circumftance which has more than once
expofed me to very difagreeable accidents, in
confequence of the burfting of the phials.

I do not know any inquiry more promifing
than the inveftigation of the properties of *nitre,*
the *nitrous acid,* and *nitrous air.* Some of the
moft wonderful phenomena in nature are con-

T nected

nected with them, and the fubject feems to be fully within our reach.

§ 2. *Speculations arifing from the confideration of the fimilarity of the* ETECTRIC MATTER *and* PHLOGISTON.

There is nothing in the hiftory of philofophy more ftriking than the rapid progrefs of *electricity.* Nothing ever appeared more trifling than the firft effects which were obferved of this agent in nature, as the attraction and repulfion of ftraws, and other light fubftances. It excited more attention by the flafhes of *light* which it exhibited. We were more ferioufly alarmed at the electrical *fhock,* and the effects of the electrical *battery* ; and we were aftonifhed to the higheft degree by the difcovery of the fimilarity of electricity with *lightning,* and the *aurora borealis,* with the connexion it feems to have with *water-fpouts, hurricanes,* and *earthquakes,* and alfo with the part that is probably affigned to it in the fyftem of *vegetation,* and other the moft important proceffes in nature.

Yet, notwithftanding all this, we have been, within a few years, more puzzled than ever with the electricity of the *torpedo,* and of the *anguille temblante* of Surinam, efpecially fince that moft curious difcovery of Mr. Walfh's,

that

that the former of thefe wonderful fifhes has the power of giving a proper electrical fhock ; the electrical matter which proceeds from it performing a real circuit from one part of the animal to the other ; while both the fifh which performs this experiment and all its apparatus are plunged in water, which is known to be a conducting fubftance.

Perhaps, however, by confidering this fact in connexion with a few others, and efpecially with what I have lately obferved concerning the identity of electricity and phlogifton, a little light may be thrown upon this fubject, in confequence of which we may be led to confider electricity in a ftill more important light. Many of my readers, I am aware, will fmile at what I am going to advance ; but the apprehenfion of this fhall not interrupt my fpeculations, how chimerical foever they may be thought to be.

The facts, the confideration of which I would combine with that of the electricity of the torpedo, are the following.

Firft, The remarkable electricity of the feathers of a paroquet, obferved by Mr. Hartmann, an account of which may be feen in Mr. Rozier's Journal for Sept. 1771. p. 69. This

T 2

bird

bird never drinks, but often wafhes itfelf; but the perfon who attended it having neglected to fupply it with water for this purpofe, its feathers appeared to be endued with a proper electrical virtue, repelling one another, and retaining their electricity a long time after they were plucked from the body of the bird, juft as they would have done if they had received electricity from an excited glafs tube.

Secondly, The electric matter directed through the body of any mufcle forces it to contract. This is known to all perfons who attend to what is called the electrical fhock; which certainly occafions a proper *convulfion*, but has been more fully illuftrated by Father Beccaria. See my *Hiftory of Electricity*, p. 402.

Laftly, Let it be confidered that the proper nourifhment of an animal body, from which the fource and materials of all mufcular motion muft be derived, is probably fome modification of phlogifton. Nothing will nourifh that does not contain phlogifton, and probably in fuch a ftate as to be eafily feparated from it by the animal functions.

That the fource of mufcular motion is phlogifton is ftill more probable, from the confideration of the well known effects of vinous and
fpirituous

fpirituous liquors, which confift very much of
phlogifton, and which inftantly brace and
ftrengthen the whole nervous and mufcular
fyftem ; the phlogifton in this cafe being, perhaps, more eafily extricated, and by a lefs
tedious animal procefs, than in the ufual method of extracting it from mild aliments.
Since, however, the mildeft aliments do the
fame thing more flowly and permanently, that
fpirituous liquors do fuddenly and tranfiently,
it feems probable that their operation is ulti-
mately the fame.

This conjecture is likewife favoured by my
obfervation, that refpiration and putrefaction affect common air in the fame manner, and in
the fame manner in which all other procefles
diminifh air and make it noxious, and which
agree in nothing but the emiffion of phlogifton.
If this be the cafe, it fhould feem that the
phlogifton which we take in with our aliment,
after having difcharged its proper function in
the animal fyftem (by which it probably undergoes fome unknown alteration) is difcharged as
effete by the lungs into the great common *menftruum*, the atmofphere.

My conjecture fuggefted (whether fupported
or not) by thefe facts, is, that animals have a
power of cÒnverting phlogifton, from the ftate

T 3   in

in which they receive it in their nutriment, into
that ſtate in which it is called the electrical
fluid; that the brain. beſides its other proper
uſes, is the great laboratory and repoſitory for
this purpoſe; that by means of the nerves this
great principle, thus exalted, is directed into
the muſcles, and forces them to act, in the
ſame manner as they are forced into action
when the electric fluid is thrown into them *ab
extra.*

I farther ſuppoſe, that the generality of animals have no power of throwing this generated electricity any farther than the limits of
their own ſyſtem; but that the *torpedo*, and
animals of a ſimilar conſtruction, have likewiſe the power, by means of an additional apparatus, of throwing it farther, ſo as to affect
other animals, and other ſubſtances at a diſtance
from them.

In this caſe, it ſhould ſeem that the electric
matter diſcharged from the animal ſyſtem (by
which it is probably more exhauſted and fa-
tigued than by ordinary muſcular motion) would
never return to it, at leaſt ſo as to be capable
of being made uſe of a ſecond time, and yet
if the ſtructure of theſe animals be ſuch
as that the electric matter ſhall dart from
one part of them only, while another part
                                                    is

is left fuddenly deprived of it, it may make a circuit, as in the Leyden phial.

As to the *manner* in which the electric matter makes a mufcle contract, I do not pretend to have any conjecture worth mentioning. I only imagine that whatever can make the muf-cular fibres recede from one another farther than the parts of which they confift, muft have this effect.

Poffibly, the *light* which is faid to proceed from fome animals, as from cats and wild beafts, when they are in purfuit of their prey in the night, may not only arife, as it has hitherto been fuppofed to do, from the friction of their hairs or briftles, &c. but that violent mufcular exertion may contribute to it. This may affift them occafionally to catch their prey; as glow-worms, and other infects, are provided with a conftant light for that purpofe, to the fupply of which light their nutriment may alfo contribute.

I would not even fay that the light which is faid to have proceeded from fome human bodies, of a particular temperament, and efpe-cially on fome extraordinary occafions, may not have been of the electrical kind, that is, produced independently of friction, or with lefs

T 4   friction

friction than would have produced it in other
perfons ; as in thofe cafes related by Bartholin
in his treatice *De luce animalium.* See particularly what he fays concerning Theodore king of
the Goths, p. 54, concerning Gonzaga duke of
Mantua, p. 57, and Gothofred Antonius, p.
123. But I would not have my readers fuppofe
that I lay much ftrefs upon ftories no better
authenticated than thefe.

The electric matter in paffing through non-
conducting fubftances always emits *light.* This
light I have been fometimes inclined to fufpect
might have been fupplied from the fubftance
through which it paffes. But I find that after
the electric fpark has diminifhed a quantity of
air as much as it poffibly can, fo that it has
no more vifible effect upon it, the electric light
in that air is not at all leffened. It is probable, therefore, that electric light comes from
the electric matter itfelf ; and this being a modification of phlogifton, it is probable that *all
light* is a modification of phlogifton alfo. Indeed,
fince no other fubftances befides fuch as contain
phlogifton are capable of ignition, and confequently of becoming luminous, it was on this
account pretty evident, prior to thefe deductions from electrical phenomena, that light and
phlogifton are the fame thing, in different
forms or ftates.

It

It appears to me that *heat* has no more proper connexion with phlogifton than it has with water, or any other conftituent part of bodies; but that it is a ftate into which the parts of bodies are thrown by their action and reaction with refpect to one another; and probably (as the Englifh philofophers in general have fuppofed) the heated ftate of bodies may confift of a fubtle vibratory motion of their parts. Since the particles which conftitute light are thrown from luminous bodies with fuch amazing velocity, it is evident that, whatever be the caufe of fuch a projection, the reaction confequent upon it muft be confiderable. This may be fufficient not only to keep up, but alfo to increafe the vibration of the parts of thofe bodies in which the phlogifton is not very firmly combined; and the difference betwèen the fubftances which are called *inflammable* and others which alfo contain phlogifton may be this, that in the former the heat, or the vibration occafioned by the emiffion of their own phlogifton, may be fufficient to occafion the emiffion of more, till the whole be exhaufted; that is, till the body be reduced to afhes. Whereas in bodies which are not inflammable, the heat occafioned by the emiffion of their own phlogifton may not be fufficient for this purpofe, but an additional heat *ab extra* may be neceffary.

Some

Some philofophers diflike the term *phlogifton*; but, for my part, I can fee no objection to giving that, or any other name, to a *real fomething*, the prefence or abfence of which makes fo remarkable difference in bodies, as that of *metallic calces* and *metals*, *oil of vitriol* and *brimftone*, &c. and which may be transferred from one fubftance to another, according to certain known laws, that is, in certain definite circumftances. It is certainly hard to conceive how any thing that anfwers this defcription can be only a mere *quality*, or mode of bodies, and not a *fubftance* itfelf, though incapable of being exhibited alone. At leaft, there can be no harm in giving this name to any *thing*, or any *circumftance* that is capable of producing thefe effects. If it fhould hereafter appear not to be a fubftance, we may change our phrafeology, if we think proper.

On the other hand I diflike the ufe of the term *fire*, as a conftituent principle of natural bodies, becaufe, in confequence of the ufe that has generally been made of that term, it includes another thing or circumftance, viz. *heat*, and thereby becomes ambiguous, and is in danger of mifleading us. When I ufe the term phlogifton, as a principle in the conftitution of bodies, I cannot miflead myfelf or others, becaufe I ufe one and the fame term to denote

denote only one and the fame *unknown caufe* of certain well-known effects. But if I fay that *fire* is a principle in the conftitution of bodies, I muft, at leaft, embarrafs myfelf with the diftinction of fire *in a ftate of action*, and fire *inactive*, or quiefcent. Befides I think the term phlogifton preferable to that of fire, becaufe it is not in common ufe, but confined to philofophy ; fo that the ufe of it may be more accurately afcertained.

Befides, if phlogifton and the electric matter be the fame thing, though it cannot be exhibited alone, in a *quiefcent ftate*, it may be exhibited alone under one of its modifications, when it is in *motion*. And if light be alfo phlogifton, or fome modification or fubdivifion of phlogifton, the fame thing is capable of being exhibited alone in this other form alfo.

In my paper on the *conducting power of charcoal*, (See Philofophical Tranfactions, vol. 60. p. 221) I obferved that there is a remarkable refemblance between metals and charcoal ; as in both thefe fubftances there is an intimate union of phlogifton with an earthy bafe ; and I faid that, had there been any phlogifton in *water*, I fhould have concluded, that there had been no conducting power in nature, but in confequence of an union of this principle with
fome

fome bafe ; for while metals have phlogifton
they conduct electricity, but when they are deprived of it they conduct no lónger. Now the
affinity which I have obferved between phlogifton and water leads me to conclude that water, in its natural ftate, does contain fome portion of phlogifton ; and according to the hypothefis juft now mentioned they muft be intimately united, becaufe water is not inflammable.

I think, therefore, that after this ftate of
hefitation and fufpence, I may venture to lay
it down as a characteriftic diftinction between
conducting and non-conducting fubftances, that
the former contain phlogifton intimately united
with fome bafe, and that the latter, if they contain phlogifton at all, retain it more loofely. In
what manner this circumftance facilitates the
paffing of the electric matter through one fubftance, and obftructs its paffage through another, I do not pretend to fay. But it is no inconfiderable thing to have advanced but *one ftep*
nearer to an explanation of fo very capital a
diftinction of natural bodies, as that into conductors and non-conductors of electricity.

I beg leave to mention in this place, as favourable to this hypothefis, a moft curious difcovery made very lately by Mr. Walfh, who
being

being affifted by Mr. De Luc to make a more
perfect vacuum in the double or arched baro-
meter, by boiling the quickfilver in the tube,
found that the electric fpark or fhock would
no more pafs through it, than through a ftick
of folid glafs. He has alfo noted feveral circumftances that affect this vacuum in a very
extraordinary manner. But fuppofing that vacuum to be perfect, I do not fee how we can
avoid inferring from the fact, that fome *fubftance* is neceffary to conduct electricity ; and
that it is not capable, by its own expanfive
power, of extending itfelf into fpaces void of
all matter, as has generally been fuppofed, on
the idea of there being nothing to obftruct its
paffage.

Indeed if this was the cafe, I do not fee how
the electric matter could be retained within the
body of the earth, or any of the planets, or
folid orbs of any kind. In nature we fee it make
the moft fplendid appearance in the upper and
thinner regions of the atmofphere, juft as it
does in a glafs tube nearly exhaufted ; but if
it could expand itfelf beyond that degree of ra-
rity, it would neceffarily be diffufed into the
furrounding vacuum, and continue and be condenfed there, at leaft in a greater proportion
than in or near any folid body, as Newton fuppofed concerning his *ether.*

If

If that mode of vibration which conftitutes heat be the means of converting phlogifton from that ftate in which it makes a part of folid bodies, and eminently contributes to the firmnefs of their texture into that ftate in which it diminifhes common air; may not that peculiar kind of vibration by which Dr. Hartley fuppofes the brain to be affected, and by which he endeavours to explain all the phenomena of fenfation, ideas, and mufcular motion, be the means by which the phlogifton, which is conveyed into the fyftem by nutriment, is converted into that form or modification of it of which the electric fluid confifts.

Thefe two ftates of phlogifton may be conceived to refemble, in fome meafure, the two ftates of fixed air, viz. elaftic, or non-elaftic; a folid, or a fluid.

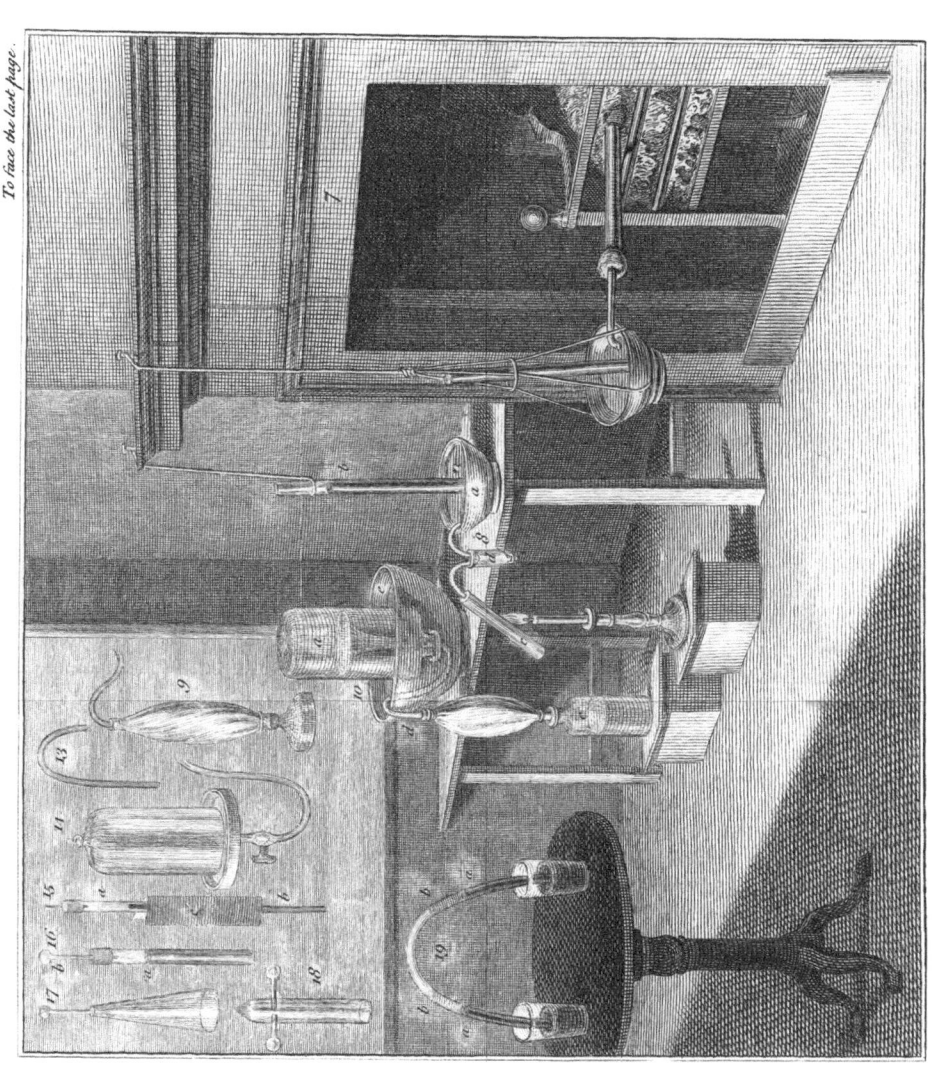

The material originally positioned here is too large for reproduction in this reissue. A PDF can be downloaded from the web address given on page iv of this book, by clicking on 'Resources Available'.

THE

APPENDIX.

IN this Appendix I shall present the reader with the communications of several of my friends on the subject of the preceding work. Among them I should with pleasure have inserted some curious experiments, made by Dr. Hulme of Halifax, on the air extracted from Buxton water, and on the impregnation of several fluids, with different kinds of air; but that he informs me he proposes to make a separate publication on the subject.

NUM-

N U M B E R   I.

EXPERIMENTS *made by Mr. Hey to prove that there is no* OIL *of* VITRIOL *in water impregnated with* FIXED AIR,

It having been fuggefted, that air arifing from a fermenting mixture of chalk and oil of vitriol might carry up with it a fmall portion of the vitriolic acid, rendered volatile by the act of fermentation ; I made the following experiments, in order to difcover whether the acidulous tafte, which water impregnated with fuch air affords, was owing to the pre- fence of any acid, or only to the fixed air it had abforbed.

EXPERIMENT I.

I mixed a tea-fpoonful of fyrup of violets with an ounce of diftilled water, faturated with fixed air procured from chalk by means of the vitriolic acid; but neither upon the firft mixture, nor after ftanding 24 hours, was the colour of the fyrup at all changed, except by its fimple dilution.

EXPERIMENT II.

A portion of the fame diftilled water, unimpregnated with fixed air, was mixed with the fyrup in the fame proportion : not the leaft difference in colour could be perceived betwixt this and the above-mentioned mixture.

EXPE-

EXPERIMENT III.

One drop of oil of vitriol being mixed with a pint of the fame diftilled water, an ounce of this water was mixed with a tea-fpoonful of the fyrup. This mixture was very diftinguifhable in colour from the two former, having a purplifh caft, which the others wanted.

EXPERIMENT IV.

The diftilled water impregnated with fo fmall a quantity of vitriolic acid, having a more agreeable tafte than when alone, and yet manifefting the prefence of an acid by means of the fyrup of violets ; I fubjeſted it to fome other tefts of acidity. It formed curds when agitated with foap, lathered with difficulty, and very imperfeſtly ; but not the leaft ebullition could be difcovered upon dropping in fpirit of fal ammoniac, or folution of falt of tartar, though I had taken care to render the latter free from caufticity by impregnating it with fixed air.

EXPERIMENT V.

The diftilled water faturated with fixed air neither effervefced, nor fhewed any clouds, when mixed with the fixed or volatile alkali.

EXPERIMENT VI.

No curd was formed by pouring this water upon an equal quantity of milk, and boiling them together.

U EXPE-

EXPERIMENT VII.

When agitated with foap, this water produced curds, and lathered with fome difficulty ; but not fo much as the diftilled water mixed with vitriolic acid in the very fmall proportion above-mentioned. The fame diftilled water without any impregnation of fixed air lathered with foap without the leaft previous curdling. River water, and a pleafant pump-water not remarkably hard, were compared with thefe. The former produced curds before it lathered, but not quite in fo great a quantity as the diftilled water impregnated with fixed air: the latter caufed a ftronger curd than any of the others above-mentioned.

EXPERIMENT VIII.

Apprehending that the fixed air in the diftilled water occafioned the coagulation, or feparation of the oily part of the foap, only by deftroying the caufticity of the *lixivium,* and thereby rendering the union lefs perfect betwixt that and the tallow, and not by the prefence of any acid; I impregnated a frefh quantity of the fame diftilled water with fixed air, which had paffed through half a yard of a wide barometer-tube filled with falt of tartar ; but this water caufed the fame curdling with foap as the former had done, and appeared in every refpect to be exactly the fame.

Experiment IX.

Diftilled water faturated with fixed air formed a white cloud and precipitation, upon being mixed with a folution of *faccharum faturni*. I found likewife, that fixed air, after paffing through the tube filled with alkaline falt, upon being let into a phial containing a folution of the metalic falt in diftilled water, caufed a perfect feparation of the lead, in the form of a white powder; for the water, after this precipitation, fhewed no cloudinefs upon a frefh mixture of the fubftances which had before rendered it opaque.

NUM.

NUMBER II.

A Letter from Mr. HEY *to* Dr. PRIESTLEY, *concerning the Effects of fixed Air applied by way of Clyster.*

Leeds, Feb. 15th, 1772.

Reverend Sir,

Having lately experienced the good effects of fixed air in a putrid fever, applied in a manner, I believe not heretofore made use of, I thought it proper to inform you of the agreeable event, as the method of applying this powerful corrector of putrefaction took its rise principally from your obfervations and experiments on factitious air ; and now, at your requeft, I fend the particulars of the cafe I mentioned to you, as far as concerns the adminiftration of this remedy.

January 8, 1772, Mr. Lightbowne, a young gentleman who lives with me, was feized with a fever, which, after continuing about ten days, began to be attended with thofe fymptoms that indicate a putrefcent ftate of the fluids.

18th, His tongue was black in the morning when I firft vifited him, but the blacknefs went off in the day-time upon drinking: He had begun to doze much the preceding day, and now he took little notice of thofe that were about him: His belly was loofe, and had been fo for fome days: his pulfe beat 110 ftrokes in a minute, and was rather low : he was ordered to take twenty-five grains of Peruvian bark with five of

tor-

tormentil-root in powder every four hours, and to
ufe red wine and water cold as his common drink.

19th, I was called to vifit him early in the morning, on account of a bleeding at the nofe which had
come on: he loft about eight ounces of blood, which
was of a loofe texture: the hæmorrhage was fup-
preffed, though not without fome difficulty, by
means of tents made of foft lint, dipped in cold water ftrongly impregnated with tincture of iron, which
were introduced within the noftrils quite through to
their pofterior apertures ; a method which has never
yet failed me in like cafes. His tongue was now
covered with a thick black pellicle, which was not diminifhed by drinking: his teeth were furred with the
fame kind of fordid matter, and even the roof of his
mouth and fauces were not free from it: his loofenefs
and ftupor continued, and he was almoft inceffantly
muttering to himfelf: he took this day a fcruple of
the Peruvian bark with ten grains of tormentil every
two or three hours: a ftarch clyfter, containing a
drachm of the compound powder of bole, without
opium, was given morning and evening: a window
was fet open in his room, though it was a fevere
froft, and the floor was frequently fprinkled with
vinegar.

20th, He continued nearly in the fame ftate : when
roufed from his dozing, he generally gave a fenfible
anfwer to the queftions afked him; but he immediately relapfed, and repeated his muttering. His fkin was
dry, and harfh, but without *petechiæ*. He fometimes
voided his urine and *fæces* into the bed, but generally
had fenfe enough to afk for the bed-pan: as he now

naufeates

naufeated the bark in fubftance, it was exchanged for
Huxham's tincture, of which he took a table fpoon-
ful every two hours in a cup full of cold water: he
drank fometimes a little of the tincture of rofes, but
his common liquors were red wine and water, or rice-
water and brandy acidulated with elixir of vitriol:
before drinking, he was commonly requefted to rinfe
his mouth with water to which a little honey and vinegar had been added. His loofenefs rather increafed,
and the ftools were watery, black, and fœtid: It was
judged neceffary to moderate this difcharge, which
feemed to fink him, by mixing a drachm of the *the-
riaca Andromachi* with each clyfter.

21ft. The fame putrid fymptoms remained, and a
fubfultus tendinum came on: his ftools were more fœtid;
and fo hot, that the nurfe affured me fhe could not apply her hand to the bed-pan, immediately after they
were difcharged, without feeling pain on this account: The medicine and clyfters were repeated.

Reflecting upon the difagreeable neceffity we feemed
to lie under of confining this putrid matter in the inteftines, left the evacuation fhould deftroy the *vis vitæ*
before there was time to correct its bad quality, and
overcome its bad effects, by the means we were ufing;
I confidered, that, if this putrid ferment could be
more immediately corrected, a ftop would probably
be put to the flux, which feemed to arife from, or at
leaft to be encreafed by it; and the *fomes* of the difeafe would likewife be in a great meafure removed. I
thought nothing was fo likely to effect this, as the in-
troduction of fixed air into the alimentary canal,
which, from the experiments of Dr. Macbride, and
thofe

thofe you have made fince his publication, appears to
be the moft powerful corrector of putrefaction hitherto known. I recollected what you had recommended
to me as deferving to be tried in putrid difeafes, I
mean, the injection of this kind of air by way of
clyfter, and judged that in the prefent cafe fuch a method was clearly indicated.

The next morning I mentioned my reflections to
Dr. Hird and Dr. Crowther, who kindly attended
this young gentleman at my requeft, and propofed
the following method of treatment, which, with
their approbation, was immediately entered upon.
We firft gave him five grains of ipecacuanha, to eva-
cuate in the moft eafy manner part of the putrid *colluvies :* he was then allowed to drink freely of brifk
orange-wine, which contained a good deal of fixed
air, yet had not loft its fweetnefs. The tincture of
bark was continued as before; and the water which
he drank along with it, was impregnated with fixed
air from the atmofphere of a large vat of fermenting
wort, in the manner I had learned from you. Inftead
of the aftringent clyfter, air alone was injected, collected from a fermenting mixture of chalk and oil of
vitriol : he drank a bottle of orange-wine in the courfe
of this day, but refufed any other liquor except water and his medicine : two bladders full of air were
thrown up in the afternoon.

23d. His ftools were lefs frequent ; their heat likewife and peculiar *fœtor* were confiderably diminifhed;
his muttering was much abated, and the *fubfultus tendinum* had left him. Finding that part of the air was
rejected when given with a bladder in the ufual way, I

eontrived a method of injecting it which was not fo liable to this inconvenience. I took the flexible tube of that inftrument which is ufed for throwing up the fume of tobacco, and tied a fmall bladder to the end of it that is connected with the box made for receiving the tobacco, which I had previoufly taken off from the tube : I then put fome bits of chalk into a fix ounce phial until it was half filled ; upon thefe I poured fuch a quantity of oil of vitriol as I thought capable of faturating the chalk, and immediately tied the bladder, which I had fixed to the tube, round the neck of the phial : the clyfter-pipe, which was faftened to the other end of the tube, was introduced into the *anus* before the oil of vitriol was poured upon the chalk. By this method the air paffed gradually into the inteftines as it was generated ; the rejection of it was in a great meafure prevented ; and the inconvenience of keeping the patient uncovered during the operation was avoided.

24th, He was fo much better, that there feemed to be no neceffity for repeating the clyfters : the other means were continued. The window of his room was now kept fhut.

25th, All the fymptoms of putrefcency had left him ; his tongue and teeth were clean; there remained no unnatural blacknefs or *fœtor* in his ftools, which had now regained their proper confiftence ; his dozing and muttering were gone off; and the difagreeable odour of his breath and perfpiration was no longer perceived. He took nourifhment to-day, with pleafure ;

fure; and, in the afternoon, fat up an hour in his chair.

His fever, however, did not immediately leave him; but this we attributed to his having caught cold from being incautioufly uncovered, when the window was open, and the weather extremely fevere; for a cough, which had troubled him in fome degree from the beginning, increafed, and he became likewife very hoarfe for feveral days, his pulfe, at the fame time, growing quicker: but thefe complaints alfo went off, and he recovered, without any return of the bad fymptoms above-mentioned.

I am, Reverend Sir,

Your obliged humble Servant,

WM. HEY.

P O S T C R I P T.

October 29, 1772.

Fevers of the putrid kind have been fo rare in this town, and in its neighbourhood, fince the commencement of the prefent year, that I have not had an opportunity of trying again the effects of fixed air, given by way of clyfter, in any cafe exactly fimilar to Mr. Lightbowne's. I have twice given water faturated with fixed air in a fever of the putrefcent kind, and it agreed very well with the patients. To one of them the aerial clyfters were adminiftred, on account of a

loofe-

loofenefs, which attended the fever, though the ftools
were not black, nor remarkably hot or fœtid.

Thefe clyfters did not remove the loofenefs, though
there was often a greater interval than ufual betwixt
the evacuations, after the injection of them. The
patient never complained of any uneafy diftention of
the belly from the air thrown up, which, indeed, is
not to be wondered at, confidering how readily this
kind of air is abforbed by aqueous and other fluids, for
which fufficient time was given, by the gradual manner of injecting it. Both thofe patients recovered
though the ufe of fixed air did not produce a crifis before the period at which fuch fevers ufually terminate.
They had neither of them the opportunity of drinking
fuch wine as Mr. Lightbowne took, after the ufe of
fixed air was entered upon ; and this, probably, was
fome difadvantage to them.

I find the methods of procuring fixed air, and impregnating water with it, which you have publifhed,
are preferable to thofe I made ufe of in Mr. Lightbowne's cafe.

The flexible tube ufed for conveying the fume of
tobacco into the inteftines, I find to be a very convenient inftrument in this cafe, by the method before-
mentioned (only adding water to the chalk, before the
oil of vitriol is inftilled, as you direct) the injection
of air may be continued at pleafure, without any other
inconvenience to the patient, than what may arife
from his continuing in one pofition during the operation, which fcarcely deferves to be mentioned, or
from

from the continuance of the clyfter-pipe within the anus, which is but trifling, if it be not fhaken much, or pufhed againft the rectum.

When I faid in my letter, that fixed air appeared to be the greateft corrector of putrefaction hitherto known, your philofophical refearches had not then made you acquainted with that moft remarkably anti-feptic property of nitrous air. Since you favoured me with a view of fome aftonifhing proofs of this, I have conceived hopes, that this kind of air may likewife be applied medicinally to great advantage.

W. H.

NUMBER III.

Obfervations on the MEDICINAL USES *of* FIXED
AIR. *By* THOMAS PERCIVAL, *M.'D. Fellow
of the* ROYAL SOCIETY, *and of the* SOCIETY
of ANTIQUARIES *in* LONDON.

Thefe Obfervations on the MEDICINAL USES OF
FIXED AIR have been before publifhed in the Second Volume of my Effays; but are here reprinted
with confiderable additions. They form a part of
an experimental inquiry into this interefting and
curious branch of Phyfics; in which the friend-
fhip of Dr. Prieftley firft engaged me, in concert
with himfelf.

Manchefter, March 16, 1774.

IN a courfe of Experiments, which is yet unfinifhed,
I have had frequent opportunities of obferving
that FIXED AIR may in no inconfiderable quantity be
breathed without danger or uneafinefs. And it is a
confirmation of this conclufion, that at Bath, where
the waters copioufly exhale this mineral fpirit,* the
bathers infpire it with impunity. At Buxton alfo,
where the Bath is in a clofe vault, the effects of fuch
effluvia, if noxious, muft certainly be perceived.

* See Dr. Falconer's very ufeful and ingenious treatife] on the Bath
water, 2d edit. p. 313.

Encou-

Encouraged by thefe confiderations, and ftill more
by the teftimony of a very judicious Phyſician at Staf-
ford, in favour of this powerful antifeptic remedy, I
have adminiſtered fixed air in a confiderable number of
cafes of the PHTHISIS PULMONALIS, by directing my
patients to infpire the ſteams of an effervefcing mixture
of chalk and vinegar ; or what I have lately preferred,
of vinegar and potaſh. The hectic fever has in feveral inſtances been confiderably abated, and the matter
expectorated has become lefs offenſive, and better di-
geſted. I have not yet been fo fortunate in any one
cafe, as to effect a cure; although the ufe of mephitic air has been accompanied with proper internal
medicines. But Dr. Withering, the gentleman referred to above, informs me, that he has been more
fuccefsful. One Phthiſical patient under his care has
by a fimilar courfe intirely recovered ; another was
rendered much better; and a third, whofe cafe was
truly deplorable, feemed to be kept alive by it more
than two months. It may be proper to obferve that
fixed air can only be employed with any profpect of
fuccefs, in the latter ſtages of the *phthiſis pulmonalis,*
when a purulent expectoration takes place. After the
rupture and difcharge of a VOMICA alfo, fuch a remedy promifes to be a powerful palliative. Antifeptic
fumigations and vapours have been long employed,
and much extolled in cafes of this kind. I made the
following experiment, to determine whether their efficacy, in any degree, depends on the feparation of
fixed air from their fubſtance.

　　　　end of a bent tube was fixed in a phial full of
lime-water ; the other end in a bottle of the tincture of
　　　　　　　　　　　　　　　　　　　　　myrrh

myrrh. The junctures were carefully luted, and the phial containing the tincture of myrrh was placed in water, heated almost to the boiling point, by the lamp of a tea-kettle. A number of air-bubbles were separated, but probably not of the mephitic kind, for no precipitation ensued in the lime water. This experiment was repeated with the *tinct. tolutanæ, ph. ed.* and with *sp, vinos. camp.* and the result was entirely the same. The medicinal action therefore of the vapours raised from such tinctures, cannot be ascribed to the extrication of fixed air; of which it is probable bodies are deprived by *chemical solution* as well as by *mixture.*

If mephitic air be thus capable of correcting purulent matter in the lungs, we may reasonably infer it will be equally useful when applied externally to foul ULCERS. And experience confirms the conclusion. Even the sanies of a CANCER, when the carrot poultice failed, has been sweetened by it, the pain mitigated, and a better digestion produced. The cases I refer to are now in the Manchester infirmary, under the direction of my friend Mr. White, whose skill as a surgeon, and abilities as a writer are well known to the public.

Two months have elapsed since these observations were written,*and the same remedy, during that period, has been assiduously applied, but without any further success. The progress of the cancers seems to be checked by the fixed air; but it is to be feared that a cure will not be effected. A palliative remedy, how-

* May, 1772.

ever,

ever, in a difeafe fo defperate and loathfome, may be
confidered as a very valuable acquifition. Perhaps NI-
TROUS AIR might be ftill more efficacious. This fpecies of factitious air is obtained from all the metals
except zinc, by means of nitrous acid; and Dr.
Prieftley informs me, that as a fweetener and antifeptic
it far furpaffes fixed air. He put two mice into a quantity of it, one juft killed, the other offenfively putrid.
After twenty-five days they were both perfectly
fweet.

In the ULCEROUS SORE THROAT much advantage
has been experienced from the vapours of effervefcing
mixtures drawn into the *fauces* †. But this remedy
fhould not fuperfede the ufe of other antifeptic applications. §

A phyfician* who had a very painful APTHOUS UL-
CER at the point of his tongue, found great relief,
when other remedies failed, from the application of
fixed air to the part affected. He held his tongue over
an effervefcing mixture of potafh and vinegar; and as
the pain was always mitigated, and generally removed
by this vaporifation, he repeated it, whenever the an-
guifh arifing from the ulcer was more than ufually fevere. He tried a combination of potafh and oil of vitriol well diluted with water; but this proved ftimu-
lant and increafed his pain; probably owing to fome

† Vid. Mr. White's ufeful treatife on the management of pregnant and
lying-in women, p. 279.
§ See the author's obfervations on the efficacy of external applications in
the ulcerous fore throats, Effays medical and experimental, Vol. I. 2d edit.
p. 377.
* The author of thefe obfervations.

particles

particles of the acid.thrown upon the tongue, by the violence of the effervefcence. For a paper ftained with the purple juice of radifhes, when held at an equal diftance over two veffels, the one containing potafh and vinegar, the other the fame alkali and *Spiritus vitrioli tenuis*, was unchanged by the former, but was fpotted with red, in various parts, by the latter.

In MALIGNANT FEVERS wines abounding with fixed air may be adminiftered, to check the feptic ferment, and fweeten the putrid *colluvies* in the *prima via*. If the laxative quality of fuch liquors be thought an objection to the ufe of them, wines of a greater age may be given, impregnated with mephitic air, by a fimple but ingenious contrivance of my friend Dr. Prieftley. †

The patient's common drink might alfo be medicated in the fame way. A putrid DIARRHÆA frequently occurs in the latter ftage of fuch diforders and it is a moft alarming and dangerous fymptom. If the difcharge be ftopped by aftringents, a putrid *fomes* is retained in the body, which aggravates the deflrium and increafes the fever. On the contrary, if it be fuffered to take its courfe, the ftrength of the patient muft foon be exhaufted, and death unavoidably enfue. The injection of mephitic air into the inteftines, under thefe circumftances, bids fair to be highly ferviceable. And a cafe of this deplorable kind, has lately been communicated to me, in which the vapour of

---

† Directions for impregnating water with fixed air, in order to communicate to it the peculiar fpirit and virtues of Pyrmont water, and other mineral waterss of a fimilar nature.

chalk

chalk and oil of vitriol conveyed into the body by
the machine employed for tobacco clyfters, quickly reftrained the *diarrhœa*, corrected the heat and fœtor
of the ftools, and in two days removed every fymptom of danger *. Two fimilar inftances of the falutary effects of mephitic air, thus adminiftered, have
occurred alfo in my own practice, the hiftory of which
I fhall briefly lay before the reader. May we not
prefume that the fame remedy would be equally ufeful in the Dysentery? The experiment is at leaft
worthy of trial.

Mr. W——, aged forty-four years, corpulent, inactive, with a fhort neck, and addicted to habits of
intemperance, was attacked on the 7th of July 1772,
with fymptoms which feemed to threaten an apoplexy.
On the 8th, a bilious loofenefs fucceeded, with a
profufe hœmorrhage from the nofe. On the 9th, I
was called to his affiftance. His countenance was
bloated, his eyes heavy, his fkin hot, and his pulfe
hard, full, and oppreffed. The diarrhæa continued;
his ftools were bilious and very offenfive; and he
complained of griping pains in his bowels. He had
loft, before I faw him, by the direction of Mr. Hall,
a furgeon of eminence in Manchefter, eight ounces
of blood from the arm, which was of a lax texture; and he had taken a faline mixture every
fixth hour. The following draught was prefcribed,
and a dofe of rhubarb directed to be adminiftered at
night.

* Referring to the cafe communicated by Mr. Hey.

X   R. *Aq.*

R. *Aq. Cinnam. ten.* ℥j.
Succ. Limon. recet. ℥ ß.
Salis Nitri gr. xij. Syr. è Succo Limon. ʒj. *M. f. Hauſt.*
4tis horis ſumendus.

July 11. The *Diarrhœa* was more moderate ; his
griping pains were abated ; and he had leſs ſtupor and
dejeƈtion in his countenance. Pulſe 90, not ſo hard or
oppreſſed. As his ſtools continued to be fœtid, the doſe
of rhubarb was repeated ; and inſtead of ſimple cin-
namon-water, his draughts were prepared with an
infuſion of columbo root.

12. The *Diarrhœa* continued ; his ſtools were involuntary ; and he diſcharged in this way a quantity
of black, grumous, and fœtid blood. Pulſe hard and
quick ; ſkin hot ; tongue covered with a dark fur ;
abdomen ſwelled ; great ſtupor. Ten grains of columbo root, and fifteen of the *Gummi rubrum aſtrin-
gens* were added to each draught. Fixed air, under
the form of clyſters, was injeƈted every ſecond or
third hour ; and direƈtions were given to ſupply the
patient plentifully with water, artificially impregnated
with mephitic air. A bliſter was alſo laid between
his ſhoulders.

13. The Diarrhœa continued, with frequent difcharges of blood ; but the ſtools had now loſt their
fœtor. Pulſe 120; great flatulence in the bowels,
and fulneſs in the belly. The clyſters of fixed air
always diminiſhed the tenſion of the *Abdomen*, abated
flatulence, and made the patient more eaſy and com-
poſed for ſome time after their injeƈtion. They were
direƈted

directed to be continued, together with the medicated water. The nitre was omitted, and a fcruple of the *Confect. Damocratis* was given every fourth hour, in an infufion of columbo root.

14. The Diarrhæa was now checked. His other fymptoms continued as before. Blifters were applied to the arms; and a drachm and a half of the *Tinctura Serpentariæ* was added to each draught.

15. His pulfe was feeble, quicker, and more irregular. He dofed much; talked incoherently; and laboured under a flight degree of *Dyfpnæa*. His urine, which had hitherto affumed no remarkable appearance, now became pale. Though he difcharged wind very freely, his belly was much fwelled, except for a fhort time after the injection of the air-clyfters. The following draughts were then prefcribed.

R *Camphoræ mucilag. G. Arab. folutæ gr. viij. Infuf. Rad. Columbo ℥jfs Tinct. Serpent. ʒij Confect. Card. ℈j Syr. è Cort. Aurant ʒi m. f. Hauft. 4tis horis fumendus.*

Directions were given to foment his feet frequently with vinegar and warm water.

16. He has had no ftools fince the 14th. His *Abdomen* is tenfe. No change in the other fymptoms. The *Tinct. Serpent.* was omitted in his draughts, and an equal quantity of *Tinct. Rhæi Sp.* fubftituted in its place.

In

In the evening he had a motion to ſtool, of which
he was for the firſt time ſo ſenſible, as to give notice
to his attendants. But the diſcharge, which was conſiderable and ſlightly offenſive, conſiſted almoſt entirely of blood, both in a coagulated and in a liquid
ſtate. His medicines were therefore varied as follows :

    ℞. *Decoct. Cort. per* ℥*iſs Tinct. Cort. ejuſd.* ℨ*ij.*
    Confect. Card. ℈*j Gum. Rubr. Aſtring. gr. xv.*
    Pulv. Alumin. gr. vij. m. f. Hauſtus 4tis horis
    ſumendus.

Red Port wine was now given more freely in his
medicated water ; and his nouriſhment conſiſted of
ſago and ſalep.

In this ſtate, with very little variation, he continued
for ſeveral days ; at one time oſtive, and at another
diſcharging ſmall quantities of fæces, mixed with grumous blood. The air-clyſters were continued, and
the aſtringents omitted.

20. His urine was now of an amber colour, and
depoſited a ſlight ſediment. His pulſe was more regu-
lar, and although ſtill very quick, abated in number
ten ſtrokes in a minute. His head was leſs confuſed,
and his ſleep ſeemed to be refreſhing. No blood appeared in his ſtools, which were frequent, but ſmall
in quantity ; and his *Abdomen* was leſs tenſe than uſual.
He was extremely deaf ; but gave rational anſwers to
the few queſtions which were propoſed to him ; and
ſaid he felt no pain.

                           21. He

21. He paffed a very reftlefs night; his delirium recurred; his pulfe beat 125 ftrokes in a minute; his urine was of a deep amber colour when firft voided; but when cold affumed the appearance of cow's whey. The *Abdomen* was not very tenfe, nor had he any further difcharge of blood.

Directions were given to fhave his head, and to wafh it with a mixture of vinegar and brandy; the quantity of wine in his drink was diminifhed; and the frequent ufe of the pediluvium was enjoined. The air clyfters were difcontinued, as his ftools were not offenfive, and his *Abdomen* lefs diftended.

22. His pulfe was now fmall, irregular, and beat 130 ftrokes in a minute. The *Dyfpnœa* was greatly increafed; his fkin was hot, and bedewed with a clammy moifture; and every fymptom feemed to indicate the approach of death. In this ftate he continued till evening, when he recruited a little. The next day he had feveral flight convulfions. His urine which was voided plentifully, ftill put on the appearance of whey when cold. Cordial and antifpafmodic draughts, compofed of camphor, tincture of caftor, and *Sp. vol. aromat.* were now directed; and wine was liberally adminiftered

24. He rofe from his bed, and by the affiftance of his attendants walked acrofs the chamber. Soon after he was feized with a violent convulfion, in which he expired.

To adduce a cafe which terminated fatally as a proof of the efficacy of any medicine, recommended to the attention of the public, may perhaps appear singular; but cannot be deemed abfurd, when that remedy anfwered the purpofes for which it was intended. For in the inftance before us; fixed air was employed, not with an expectation that it would cure the fever, but to obviate the fymptoms of putrefaction, and to allay the uneafy irritation in the bowels. The difeafe was too malignant, the nervous fyftem too violently affected, and the ftrength of the patient too much exhaufted by the difcharges of blood which he fuffered, to afford hopes of recovery from the ufe of the moft powerful antifeptics.

But in the fucceeding cafe the event proved more fortunate.

Elizabeth Grundy, aged feventeen, was attacked on the 10th of December 1772, with the ufual fymptoms of a continued fever. The common method of cure was purfued; but the difeafe increafed, and foon affumed a putrid type.

On the 23d I found her in a conftant delirium, with a *fubfultus tendinum.* Her fkin was hot and dry, her tongue black, her thirft immoderate, and her ftools frequent, extremely offenfive, and for the moft part involuntary. Her pulfe beat 130 ftrokes in a minute; fhe dofed much; and was very deaf. I directed wine to be adminiftered freely; a blifter to be applied to her back; the *pediluvium* to be ufed feveral times in the day; and mephitic air to be injected under the form of a clyfter every two hours. The next day her ftools were lefs frequent, had loft their fœtor, and were no longer difcharged involuntarily; her
pulfe

pulfe was reduced to 110 ftrokes in the minute ; and her delirium was much abated. Directions were given to repeat the clyfters, and to fupply the patient liberally with wine. Thefe means were affiduoufly purfued feveral days ; and the young woman was fo recruited by the 28th, that the injections were difcontinued. She was now quite rational, and not averfe to medicine. A decoction of Peruvian bark was therefore prefcribed, by the ufe of which fhe fpeedily recovered her health.

I might add a third hiftory of a putrid difeafe, in which the mephitic air is now under trial, and which affords the ftrongeft proof both of the *antifeptic*, and of the *tonic* powers of this remedy ; but as the iffue of the cafe remains yet undetermined (though it is highly probable, alas! that it will be fatal) I fhall relate only a few particulars of it. Mafter D. a boy of about twelve years of age, endowed with an uncommon capacity, and with the moft amiable difpofitions, has laboured many months under a hectic fever, the confequence of feveral tumours in different parts of his body. Two of thefe tumours were laid open by Mr. White, and a large quantity of purulent matter was difcharged from them. The wounds were very properly treated by this fkilful furgeon, and every fuitable remedy, which my beft judgment could fuggeft, was affiduoufly adminiftered. But the matter became fanious, of a brown colour, and highly putrid. A *Diarrhæa* fucceeded ; the patient's ftools were intolerably offenfive, and voided without his knowledge. A black fur collected about his teeth ; his tongue was covered with *Aphthæ* ; and his breath was fo fœtid, as fcarcely to be endured. His ftrength was almoft ex-

X 4 haufted

haufted ; a *fubfultus tendinum* came on ; and the final
period of his fufferings feemed to be rapidly approach-
ing. As a laft, but almoft hopelefs, effort, I advifed
the injeftion of clyfters of mephitic air. Thefe foon
correfted the fœtor of the patient's ftools ; reftrained
his *Diarrhœa* ; and feemed to recruit his ftrength and
fpirits. Within the fpace of twenty-four hours his
wounds affumed a more favourable appearance ; the
matter difcharged from them became of a better colour
and confiftence ; and was no longer fo offenfive to the
fmell. The ufe of this remedy has been continued
feveral days, but is now laid afide. A large tumour
is fuddenly formed under the right ear ; fwallowing is
rendered difficult and painful ; and the patient refufes
all food and medicine. Nourifhing clyfters are di-
refted ; but it is to be feared that thefe will renew
the loofenefs, and that this amiable youth will quickly
fink under his diforder *.

The ufe of *wort* from its faccharine quality, and
difpofition to ferment, has lately been propofed as a
remedy for the SEA SCURVY. Water or other liquors,
already abounding with fixed air in a feparate ftate,
fhould feem to be better adapted to this purpofe ; as
they will more quickly correft the putrid difpofition
of the fluids, and at the fame time, by their gentle
ftimulus ‡ increafe the powers of digeftion, and give
new ftrength to the whole fyftem.

* He languifhed about a week, and then died.

‡ The vegetables which are moft effficacious in the cure of the fcurvy,
poffefs fome degree of a ftimulating power.

Dr.

Dr. Prieſtley, who ſuggeſted both the idea and the means of executing it, has under the ſanction of the College of Phyſicians, propoſed the ſcheme to the Lords of the Admiralty, who have ordered trial to be made of it, on board ſome of his Majeſty's ſhips of war. Might it not however give additional effi- cacy to this remedy, if inſtead of ſimple water, the infuſion of malt were to be employed?

I am perſuaded ſuch a medicinal drink might be preſcribed alſo with great advantage in SCROPHULOUS COMPLAINTS, when not attended with a hectic fever ; and in other diſorders in which a general acrimony prevails, and the craſis of the blood is deſtroyed. Under ſuch circumſtances, I have ſeen *vibices* which ſpread over the body, diſappear in a few days from the uſe of wort.

A gentleman who is ſubject to a ſcorbutic eruption in his face, for which he has uſed a variety of re- medies with no very beneficial effect, has lately applied the fumes of chalk and oil of vitriol to the parts affect- ed. The operation occaſions great itching and prick- ing in the ſkin, and ſome degree of drowſineſs, but evidently abates the ſerous diſcharge, and diminiſhes the eruption. This patient has ſeveral ſymptoms which indicate a genuine ſcorbutic DIATHESIS ; and it is probable that fixed air, taken internally, would be an uſual medicine in this caſe.

The ſaline draughts of Riverius are ſuppoſed to owe their antiemetic effects to the air, which is ſepa- rated from the ſalt of wormwood during the act of efferveſcence. And the tonic powers of many mineral

waters

waters feem to depend on this principle. I was lately defired to vifit a lady who had moft fevere convul-
five REACHINGS. Various remedies had been adminiftered without effect, before I faw her. She earneftly
defired a draught of malt liquor, and was indulged
with half a pint of Burton beer in brifk effervefcence.
The vomitings ceafed immediately, and returned no
more. Fermenting liquors, it is well known, abound
with fixed air. To this, and to the cordial quality of
the beer, the favourable effect which it produced, may
juftly be afcribed. But I fhall exceed my defign by
enlarging further on this fubject. What has been advanced it is hoped, will fuffice to excite the attention
of phyficians to a remedy which is capable of being
applied to fo many important medicinal purpofes.

NUMBER IV.

Extract of a Letter from WILLIAM FAL-
CONER, M. D. *of* BATH.

Jan 6, 1774.

Reverend Sir,

I once obferved the fame tafte you mention (Philofophical Tranfactions, p. 156. of this Volume, p. 35.)
viz. like tar water, in fome water that I impregnated
with fixed air about three years ago. I did not then
know to what to attribute it, but your experiment
feems to clear it up. I happened to have fpent all my
acid for raifing effervefcence, and to fupply its place I
ufed a bottle of dulcified fpirit of nitre, which I
knew was greatly under-faturated with fpirit of wine ;
from

from which, as analogous to your obſervation, I imagine the effect proceeded.

As * to the coagulation of the blood of animals
by fixed Air, I fear it will ſcarce ſtand the teſt· of
experiment, as I this day gave it, I think, a fair
trial, in the following manner.

A young healthy man, at 20 years old, received a
contuſion by a fall, was inſtantly carried to a neighbouring ſurgeon, and, at my requeſt, bled in the
following manner.

I inſerted a glaſs funnel into the neck of a large
clear phial about ℥x. contents, and bled him into it to
about ℥viii. By theſe means the blood was expoſed to
the air as little a time as poſſible, as it flowed into
the bottle as it came from the orifice.

As ſoon as the quantity propoſed was drawn, the
bottle was carefully corked, and brought to me. It
was then quite fluid, nor was there the leaſt ſepara-
tion of its parts.

On the ſurface of this I conveyed ſeveral ſtreams of
fixed air (having firſt placed the bottle with the blood
in a bowl of water, heated as nearly to the human
heat as poſſible) from the mixture of the vitriolic acid
and lixiv. tartar, which I uſe preferably to other alka-

---

* This refers, to an experiment mentioned in the firſt publication of
theſe papers in the Philoſophical Tranſactions, but omitted in this volume.

lines

lines, as being (as Dr. Cullen obferves) in the mildeft
ftate, and therefore moft likely to generate moft air.

I fhook the phial often, and threw many ftreams of
air on the blood, as I have often practifed with fuccefs
for impregnating water ; but could not perceive the
fmalleft figns of coagulation, although it ftood in an
atmofphere of fixed air 20 minutes or more.   I
then uncorked the bottles, and poured off about ʒii
to which I added about 6 or 7 gtts of fpirit of vitriol, which coagulated it immediately. I fet the remains in a cold place and it coagulated, as near as
I could judge, in the fame time that blood would
have done newly drawn from the vein.

P. 82. Perhaps the circumftance of putrid vegetables yielding all fixed and no inflammable air may
be the caufes of their proving fo antifeptic, even when
putrid, as appears by Mr. Alexander's Experiments.

P. 86. Perhaps the putrid air continually exhaled
may be one caufe of the luxuriancy of plants growing
on dunghills or in very rich foils.

P. 146. Your obfervation that inflammable air confifts of the union of fome acid vapour with phlogifton,
puts me in mind of an old obfervation of Dr. Cullen,
that the oil feparated from foap by an acid was much
more inflammable than before, refembling effential
oil, and foluble in V. fp.

I have tried fixed air as an antifeptic taken in by
refpiration, but with no great fuccefs. In one cafe it
feemed to be of fervice, in two it feemed indifferent,
and in one was injurious, by exciting a cough.
                                        N U M-

NUMBER V.

Extract of a Letter from Mr. WILLIAM BEWLEY,
of GREAT MASSINGHAM, NORFOLK.

Dear Sir,   March 23, 1774.

When I firft received your paper, I happened to
have a procefs going on for the preparation of *nitrous
ether*, without diftillation.* I had heretofore always
taken for granted that the elaftic fluid generated in that
preparation was *fixed* air: but on examination I found
this combination of the nitrous acid with inflammable
fpirits, produced an elaftic fluid that had the fame general properties with the air that you unwillingly,
though very properly, in my òpinion, term *nitrous*;
as I believe it is not to be procured without employ-
ing the *nitrous* acid, either in a fimple ftate, or com-
pounded, as in *aqua regia*. I fhall fuggeft, however,
by and by fome doubts with refpect to it's title to the
appellation of *air*.

Water impregnated with your nitrous air *certainly*,
as you fufpected from it's tafte, contains the nitrous

---

* The firft account of this curious procefs was, I believe, given in the
Mem. de l'Ac. de Sc. de Paris for 1742. Though feemingly lefs volatile
than the vitriolic ether, it boils with a much fmaller degree of heat. One
day laft fummer, it boiled in the cooleft room of my houfe; as it gave me
notice by the explofion attending its driving out the cork. To fave the
bottle, and to prevent the total lofs of the liquor by evaporation, I found
myfelf obliged inftantly to carry it down to my cellar.

acid,

acid. On faturating a quantity of this water with a
fixed alcali, and then evaporating, &c. I have procured two chryftals of nitre. But the principal obfervations that have occurred to me on the fubject of
nitrous air are the following. My experiments have
been few and made by fnatches, under every difadvantage as to apparatus, &c. and with frequent inter-
ruptions; and yet I think they are to be depended upon.

My firft remark is, that nitrous air does not give
water a fenfibly acid impregnation, unlefs it comes into contact, or is mixed with a portion of common or
atmofpherical air: and my fecond, that nitrous air
principally confifts of the nitrous acid itfelf, reduced
to the ftate of a *permanent* vapour not condenfable by
cold, like other vapours, but which requires the prefence and admixture of common air to reftore it to its
primitive ftate of a liquid. I am beholden for this
idea, you will perceive, to your own very curious
difcovery of the true nature of Mr. Cavendifh's *marine*
vapour.

When I firft repeated your experiment of impregnating water with nitrous air, the water, I muft own
tafted acid; as it did in one, or perhaps two trials
afterwards; but, to my great aftonifhment, in all the
following experiments, though fome part of the factitious air, or vapour, was vifibly abforbed by the water, I could not perceive the latter to have acquired
any fenfible acidity. I at length found, however, that
I could render this fame water *very* acid, by means
only of the nitrous air already included in the phial
with it. Taking the inverted phial out of the water,
I remove my finger from the mouth of it, to admit a
little

little of the common air, and inftantly replace my
finger. The rednefs, effervefcence, and diminution
take place. Again taking off my finger, and inftantly
replacing it, more common air rufhes in, and the
fame phenomena recur. The procefs fometimes requires to be feven or eight times repeated, before the
whole of the nitrous *vapour* (as I fhall venture to call
it) is condenfed into nitrous *acid*, by the fucceffive
entrance of frefh parcels of common air after each effervefcence ; and the water becomes evidently more
and more acid after every fuch frefh admiffion of the
external air, which at length ceafes to enter, when
the whole of the vapour has been condenfed. No agitation of the water is requifite, except a gentle motion, juft fufficient to rince the fides of the phial, in
order to wafh off the condenfed vapour.

The acidity which you and I likewife, at firft)
obferved in the water agitated with nitrous air *alone*,
I account for thus. On bringing the phial to the
mouth, the common air meeting with the nitrous
vapour in the neck of the phial, condenfes it, and
impregnates the water with the acid, in the very act
of receiving it upon the tongue. On ftopping the
mouth of the phial with my tongue for a fhort time
and afterwards withdrawing it a very little, to fuffer
the common air to rufh paft it into the phial, the fenfation of acidity has been fometimes intolerable: but
taking a large gulph of the water at the fame time, it
has been found very flightly acid.--The following is
one of the methods by which I have given water a
very ftrong acid impregnation, by means of a mixture
of nitrous and common air.

Into

Into a fmall phial, containing only common air, I force a quantity of nitrous air at random, out of a bladder, and inftantly clap my finger on the mouth of the bottle. I then immerfe the neck of it into water, a fmall quantity of which I fuffer to enter, which fquirts into it with violence; and immediately replacing my finger, remove the phial. The water contained in it is already *very* acid, and it becomes more and more fo (if a fufficient quantity of nitrous air was at firft thrown in) on alternately ftopping the mouth of the phial, and opening it, as often as frefh air will enter.

Since I wrote the above, I have frequently converted a fmall portion of water in an ounce phial into a weak *Aqua fortis*, by repeated mixtures of common and nitrous air; throwing in alternately the one or the other, according to the circumftances; that is, as long as there was a fuperabundance of nitrous air, fuffering the common air to enter and condenfe it; and, when that was effected, forcing in more nitrous air from the bladder, to the common air which now predominated in the phial—and fo alternately. I have wanted leifure, and conveniences, to carry on this procefs to its *maximum*, or to execute it in a different and better manner; but from what I have done, I think we may conclude that nitrous air confifts principally of the nitrous acid, phlogifticated, or otherwife fo modified, by a previous commenftruation with metals, inflammable fpirits, &c. as to be reduced into a durably elaftic vapour : and that, in order to deprive it of its elafticity, and reftore it to its former ftate, an addition of common air is requifite, and, as I fufpect, of water likewife, or fome other fluid: as in the

courfe

course of my few trials, I have not yet been able to condense it in a perfectly dry bottle.

N U M B E R   VI.

A Letter from Dr. FRANKLIN.

Craven Street, April 10, 1774.

Dear Sir,

In compliance with your request, I have endeavoured to recollect the circumstances of the American experiments I formerly mentioned to you, of raising a flame on the surface of some waters there.

When I passed through New Jersey in 1764, I heard it several times mentioned, that by applying a lighted candle near the surface of some of their rivers, a sudden flame would catch and spread on the water, continuing to burn for near half a minute: But the accounts I received were so imperfect that I could form no guess at the cause of such an effect, and rather doubted the truth of it. I had no opportunity of seeing the experiment; but calling to see a friend who happened to be just returned home from making it himself, I learned from him the manner of it; which was to choose a shallow place, where the bottom could be reached by a walking-stick, and was muddy; the mud was first to be stirred with the stick, and when a number of small bubbles began to arise from it, the candle was applied. The flame was so sudden and so strong, that it catched his ruffle and spoiled it, as I saw. New-

Y

Jersey

Jersey having many pine-trees in different parts of it, I then imagined that something like a volatile oil of turpentine might be mixed with the waters from a pine-swamp, but this supposition did not quite satisfy me. I mentioned the fact to some philosophical friends on my return to England, but it was not much attended to. I suppose I was thought a little too credulous.

In 1765, the Reverend Dr. Chandler received a letter from Dr. Finley, President of the College in that province, relating the same experiment. It was read at the Royal Society, Nov. 21. of that year, but not printed in the Transactions ; perhaps because it was thought too strange to be true, and some ridicule might be apprehended if any member should attempt to repeat it in order to ascertain or refute it. The following is a copy of that account.

" A worthy gentleman, who lives at a few miles distance, informed me that in a certain small cove of a mill-pond, near his house, he was surprized to see the surface of the water blaze like inflamed spirits I soon after went to the place, and made the experiment with the same success. The bottom of the creek was muddy, and when stirred up, so as to cause a considerable curl on the surface, and a lighted candle held within two or three inches of it, the whole surface was in a blaze, as instantly as the vapour of warm inflammable spirits, and continued, when strongly agitated, for the space of several seconds. It was at first imagined to be peculiar to that place ; but upon trial it was soon found, that such a bottom in other places exhibited the same phenomenon. The disco-
very

very was accidentally made by one belonging to the mill.''

I have tried the experiment twice here in England, but without fuccefs. The firft was in a flow running water with a muddy bottom. The fecond in a ftagnant water at the bottom of a deep ditch. Being fome time employed in ftirring this water, I afcribed an intermitting fever, which feized me a few days after, to my breathing too much of that foul air which I ftirred up from the bottom, and which I could not avoid while I ftooped in endeavouring to kindle it.—The difcoveries you have lately made of the manner in which inflammable air is in fome cafes produced, may throw light on this experiment, and explain its fucceeding in fome cafes, and not in others. With the higheft efteem and refpect,

I am, Dear Sir,

Your moft obedient humble fervant,

B. FRANKLIN.

NUMBER VII.

Extract of a Letter from Mr. HENRY *of* Manchefter.

It is with great pleafure I hear of your intended publication *on air*, and I beg leave to communicate to you an experiment or two which I lately made.

Dr. Percival had tried, without effect, to diffolve lead in water impregnated with fixed air. I however thought

thought it probable, that the experiment might fucceed with nitrous air. Into a quantity of water impregnated with it, I put feveral pieces of fheet-lead,
and fuffered them, after agitation, to continue immerfed about two hours. A few drops of vol. tinƈure
of fulphur changed the water to a deep orange colour,
but not fo deep as when the fame tinƈure was added
to a glafs of the fame water, into which one drop of
a folution of fugar of lead had been inftilled. The
precipitates of both in the morning, were exaƈtly of
the fame kind; and the water in which the lead had
been infufed all night, being again tried by the fame
teft, gave-figns of a ftill ftronger faturnine impregnation—Whether the nitrous air aƈts as an acid on the
lead, or in the fame manner that fixed air diffolves it,
I do not pretend to determine. Syrup of violets added
to the nitrous water became of a pale red, but on
ftanding about an hour, grew of a turbid brown caft.

Though the nitrous acid is not often found, except
produced by art, yet as there is a probability that
nitre may be formed in the earth in large towns, and
indeed foffile nitre has been aƈtually found in fuch
fituations, it fhould be an additional caution againft
the ufe of leaden pumps.

I tried to diffolve mercury by the fame means, but
without fuccefs.

I am, with the moft fincere efteem,
Dear Sir,
Your obliged and obedient fervant,
THO. HENRY.

F I N I S.

ERRATA.

E R R A T A.

P. 15. l. 13. *for* it to *read* to it
p. 24. l. 20. — has —— had
p. 60. l. 22. — inflammable —— in inflammable
p. 84. l. 5. — experiments —— experiment
p. 145. l. 16. — with —— of
p. 153. l. 1. — that is —— this air
p. 199. l. 17. — ingenious —— ingenuous
p. 211. l. 23. — of —— , if
p. 243. l. 27. — diminifhing —— diminifhed
p. 272. l. 21. — feem —— feems
p. 301. l. 31. — — —— one end
p. 303. l. 5. — — —— the nitrous
p. 304. l. 21. — deflrium —— delirium
p. 306. l. 2. — recet. —— recent.
p. 308. l. 7. — per —— Peruv.
p. 313. l. 27. — ufual —— ufeful
p. 300. to 314. paffim—Diarrhæa-—— Diarrhœa
p. 316. l. 11. — remains —— remainder
p. 524. l. 15. — it —— iron.